Acoustic Techniques for Home & Studio

F. Alton Everest

*With a specially written chapter for
the guidance of the English reader
by W. Oliver*

FOULSHAM-TAB LIMITED
Slough Bucks England

Foulsham-Tab Limited
Yeovil Road Slough Bucks England

Acoustic Techniques
for Home & Studio

Copyright © 1973 and 1974 by
Tab Books and Foulsham-Tab Limited

Reproduction or publication of the
content in any manner, without the
express permission of the publisher, is
prohibited. No liability is assumed with
respect to the use of the information
herein.
Library of Congress Number 73-78198
Cat. Code No. 646

ISBN 0–7042–0100–3

Introduction Printed and Made in Great Britain by
A. Wheaton & Co., Exeter
Balance printed in U.S.A.

It is essential that the English reader should read this chapter.

If you are going to the expense or trouble of buying or building the very best equipment you can get, the last thing you will want to do is to detract from its performance by using it in any totally unsuitable environment where the acoustic properties of the listening area are all wrong!

A few extra pounds expended on improving your listening-room may well prove to be money well spent. It is equally important to make sure of optimum acoustical environment in any studio that is used for the initial production or processing of sound that is to be transmitted by radio or by cable network, or that is to be recorded on disc or tape.

Admittedly one can compensate electronically to some extent, at least, for certain defects or deficiencies due to poor acoustics. Some present-day amplifiers can offer some very special facilities to do just this. But you will reduce the need to manipulate the signal in the electronic chain between transmitting and receiving ends if you do everything possible to create favourable conditions for the initial production, or ultimate reproduction, of the actual sounds.

This book deals very thoroughly with the problems, and their solutions, which are involved in the creation of these ideal acoustic conditions. In his Preface, the American author (this book being of American origin) describes acoustic techniques as the "most neglected link in the communication chain" and says: "All too often one finds the finest microphones, amplifiers, loudspeakers and tape recorders money can buy associated with abominable room acoustics."

The discussion about sound in the first chapter presents some features that will be already familiar to the radio enthusiast—such things as sine waves and units such as wavelength or frequency. But whereas in radio we are most often concerned with waves that travel through space with the speed of light (about 186,000 miles per *second*), in the field of this book we are more concerned with very much slower-moving waves, namely sound waves, which travel through air at about 750 miles per *hour*.

The relevant formula for each of the conversions that you may need to make will be found on p. 14. Fletcher-Munson curves, giving equal-loudness contours for average human hearing, are on p. 29. These curves emphasize the lack of flatness in the response of the human ear to sounds at different frequencies. Another graph, on p. 31, shows that the frequency-response of the human ear is not flat, but is flatter for very loud sounds than for soft sounds—a characteristic for which the "loudness control" found on some amplifiers is intended to compensate.

Throughout the book there are curves, graphs and tables giving useful practical data relating to environmental acoustics, and also helping to explain some of the theoretical considerations that come into the discussion when dealing with the basic principles involved.

Materials for do-it-yourself projects in loudspeaker enclosures etc. are obtainable from various firms. One example which may be instanced as a useful source of supply is Nichols Acoustical Fitments (address at end of chapter). They are currently listing such items as acoustically-transparent black foam padding; long-haired wool fibres; special wadding; and covering materials such as terylene, vynair and tygan.

Vibrating woodwork, floors, etc., can cause trouble, even if only in upsetting the stability of a delicately balanced pick-up arm on a record player. So it pays to mount your hi-fi equipment on something really firm and solid. Special adjustable shelving is available from a range known as Click Adjustable Shelving (address at end of chapter). The fittings include brackets infinitely adjustable for height. Wiring cables can be tucked away inside hollow uprights where they are concealed neatly by snap-on cover strips. Another special item included with the system is an adjustable spotlight and reading light to illuminate your equipment and thereby make it easier to handle.

In the section on Pinpointing Room Colourations (pp. 186–187), the author mentions the use of a tunable amplifier by the BBC to accentuate a narrow frequency band. It occurs to the writer that some other equipment might be adaptable to this purpose. Though obviously not so applicable as a device expressly designed or selected to meet the need, other devices just *might* have an incidental application in this field. It is perhaps at least worth a thought, and possibly a little experimenting!

Thinking on these lines brings to mind an arrangement on one of the Eagle amplifiers, the AA6 (a 20-watt per channel model). Instead of having just the conventional bass and treble controls, this amplifier has five slide controllers which boost or cut five separate sectors (stated to be around 40, 200, 1,200, 6,000 and 15,000 Hz) within the total spectrum covered by the reproduction system. The makers claim that this arrangement enables you to bring out and clarify or emphasize voices or instruments. The effect can range from "de-muffling" speech or song to magnifying the throb of a double bass or adding a sparkling brilliance to the tone of a flute.

The manufacturers also state that this amplifier "can compensate for those poor room acoustics you always meant to do something about." While the present book aims at encouraging you to *do* something about room acoustics, an amplifier with a facility for multiple frequency-sector control could be a very useful help in overcoming any problems that prove to be otherwise obdurate; and might also be a valuable aid in detecting or identifying particular acoustic defects in the environment.

On p. 179 there is a reference to the Tektronix Type 561 oscilloscope used at the Moody Institute of Science. Many firms producing or marketing electronic equipment in the United States have subsidiaries or associated companies, or agents or distributors, on this side of the Atlantic. Sometimes the international link is obvious from the name of the company on each side of the ocean; sometimes one has to do quite a bit of research to find out who or what organization handles any particular line of goods over here. In the case of Tektronix equipment, the firm on this side of the Atlantic is Telequipment Sales, Tektronix U.K. Ltd. (address at end of this chapter).

SUPPLIERS

The various suggestions for treating the walls, ceiling, etc. of listening rooms and studios, in this book, involve the use of more or less specialized materials. Sources of supply may be found among the various "home and trade" firms catering for do-it-yourself enthusiasts, small builders, etc. There may well br such a firm in your own locality, since D-I-Y centres have sprung up all over the country of recent years.

In the present chapter we have referred to a few firms whose present address, at the time of writing, we give below. But as addresses are liable to change from time to time, it is essential to verify details frlm the latest information in advertisements etc. These may be found in the various technical journals.

NICHOLS ACOUSTICAL FITMENTS
Bubwith, Selby, Yorkshire.

CLICK ADJUSTABLE SHELVING
220, Queenstown Road, London, SW8.

EAGLE INTERNATIONAL
Heather Park Drive, Wembley, HA0 1SU.

TELEQUIPMENT SALES, TEKTRONIX U.K. LTD.
Beaverton House, P.O. Box 69, Harpenden, Herts.

The various electronic firms now marketing audio, hi-fi, stereo, quad, etc. are far too numerous to mention here, but you will find names and addresses galore in the journals catering for this field of interest.

PREFACE

How high is **hi**? How faithful to the original is **fi**? **Hi-fi** means different things to different people. Calling a pocket radio **hi-fi** shows the depths to which the term **high fidelity** has been dragged.

In this book, which covers what every person interested in hi-fi should know about acoustics, the term **hi-fi** is taken in its best and highest sense, the serious striving after quality and reality in the recording and reproduction of sound. Defined in this way "hi-fi" encompasses a wide range of persons, not only the spare-time enthusiast but the professional audio artists and engineers as well. Professional audio people on both sides of the mike in radio, television, recording and in the electronic industry have a vested interest in this most neglected link in the communication chain, **environmental acoustics.**

The inherent intangibility of acoustic problems has given rise to a tendency to sweep such things under the rug and concentrate on the more tractable electronic links. There has arisen an almost superstitious attitude toward the mysteries of acoustics. An example of this is the abuse of the so-called "acoustic tile" in sound treatment of rooms. There is a feeling that the more used, the better will be the acoustics. This is bunk. All too often one finds the finest microphones, amplifiers, loudspeakers, and tape recorders money can buy associated with abominable room acoustics.

Some understanding of the strange things a room full of air is capable of doing to audio signals is basic to an informed use of the room for recording or reproduction of sound. A room full of air is capable of vibrating in many different modes and, with a train of harmonics for each mode, gives rise to combinations of almost unbelievable complexity. Fortunately, there has emerged a very practical and relatively simple method of analyzing studios and listening rooms. By this technique, and designing for proper reverberation time, the performance of a proposed room can be predicted within limits, and defects in an existing room can be identified and treated. If properly controlled this "room effect" lends life and interest to sounds originating or reproduced in the room in contrast to the deadness of sounds outdoors. Room environment has an important bearing on sound quality in a listening system whether it be mono, stereo, or quad.

This is intended more as an understand-it-yourself than a do-it-yourself book. When building a home the need for an architect is widely recognized. It makes just as much sense to hire a qualified acoustic consultant in solving acoustic problems, especially where large investments are concerned. For the home hi-fi enthusiast and the operators of small broadcast or recording studios, often at remote locations, in foreign lands, or in dire financial straits, the consultant may be out of the question. For these, this book points the way to reasonable solutions to many common problems.

F. Alton Everest

CONTENTS

1 HOW SOUND ACTS   7

Stimulus or Sensation?—How Is Sound Transmitted?—The Sine Wave—Wavelength and Frequency—Inverse Square Law—Reflection Of Sound—Superposition Of Sound—Sound Around Corners—Absorption Of Sound

2 HUMAN HEARING   24

Ears—Golden and Tin—How We Hear—The Decibel—The Discrimination Ability Of the Ear—The Ear As An Analyzer—Binaural Hearing

3 SPEECH, MUSIC, AND NOISE   38

Speech—Music—Dynamic Range—Line Spectra—Continuous Spectra—Phase Relationships—NoiseThe Bad Kind—Noise—The Good Kind—Distortion

4 RESONANCES IN ROOMS AND OTHER THINGS   52

Resonators—Waves Or Rays?—Resonances in Listening Rooms and Small Studios

5 STANDING WAVES IN LISTENING ROOMS AND SMALL STUDIOS   58

Standing Wave Patterns—Experiments With Colorations

6 DIFFUSION OF SOUND IN SMALL ROOMS   66

Room Proportions—Nonparallel Walls—Convex Surfaces: The Poly—Plane Surfaces—Distribution Of Absorbing Materials

7 CONTROL OF INTERFERING NOISE   77

Noise Sources, and Some Solutions—Airborne Noise—Noise Carried By Structure—Noise Transmitted By Diaphragm Action—Sound-Insulating Walls—Comparison Of Wall Structures—Double Windows—Sound-Insulating Doors—Noise and Room Resonances

8 SOUND ABSORBERS   91

Porous Absorbers—Panel Absorbers—Polys: Wraparound Panels—Perforated Panel Absorbers—Slat Absorbers—Midrange Absorbers—Modules—Placement Of Materials

9 REVERBERATION AND HOW TO COMPUTE IT   108

Reverberation and Normal Modes—Reverberation Time—T_{60}—Optimum Reverberation Time—Living Room T_{60}—Treble T_{60}—Calculation of T_{60}—The Sabine

Equation—The Eyring Reverberation Formula—The
"Soft" Studio—Two Rooms Coupled Electroacoustically

10 ACOUSTIC DESIGN OF A STUDIO 135
Main Studio—Speech Studio—Control Room—General
Design Factors

11 ADJUSTABLE ACOUSTICS 154
Draperies—Portable Panels—Rotating Elements—
Hinged Panels—Variable Resonant Devices—Louvered
Panels—The Snow Adjustable Element

12 TUNING THE LISTENING ROOM 163
The Listening Problem—The Listening Chain—Acoustic
Response—Limitations Of Room ∕ Speaker
Equalization—Tone Controls—A Home Hi-Fi Equalizing
Procedure

13 EVALUATING STUDIO ACOUSTICS 173
How Experts Evaluate Studio Acoustics—Signal Sour-
ces—Amplifier-Loudspeaker—Microphone—Filter—
Graphic Level Recorder—Reverb Time: Stopwatch
Method—Evaluation Of Background Noise—The Sound
Level Meter—Flutter Echoes—Pinpointing Room
Colorations—Evaluating Speech Intelligibility

14 A PICTORIAL TOUR OF STUDIOS AROUND THE WORLD 189

APPENDIX I—SELECTED ABSORPTION COEFFICIENTS 207

APPENDIX II—EYRING'S FORMULA FOR DEAD ROOMS 211

REFERENCES 215

INDEX 220

HOW SOUND ACTS

1

If you are the impatient type or if you already have a good background in sound and the hearing mechanism, you may wish to skip Chapters 1, 2, and 3, and leaping forward, land in the middle of the acoustics of listening rooms and small studios. These first three chapters are to make sure you end this leap on your feet—and running—rather than screeching in on your haunches with trousers smoking!

STIMULUS OR SENSATION?

"If a tree falls in a forest with no ear to hear it, is any sound produced?" This ancient question is as modern as tomorrow because it brings us face to face with the dual nature of sound. Sound may be defined as a wave motion in air or other elastic medium (**stimulus**), or as that excitation of the hearing mechanism which results in the perception of sound (**sensation**). Which definition applies in a given situation depends upon whether our approach is physical or psychophysical. Viewed in this way the ancient question can be readily answered. There would certainly be the physical stimulus as the tree falls, but if no ear is present there could be no sensation. The type of problem we are facing would thus dictate our approach to sound. If we are interested in the disturbance in air created by a loudspeaker, we approach it as a problem in physics. If we are interested in how it sounds to a person nearby we must involve psychophysical methods. In this book we are concerned with acoustics in relation to people and must therefore treat both aspects of sound.[1]

We can break these two views of sound down into terms which are more a part of the experience of the hi-fi enthusiast:

---

*References are listed beginning on page 215.

Physical Quantity	Comparable Psychophysical Quantity
Frequency	Pitch
Intensity	Loudness
Waveform	Quality (or timbre)

Frequency is a nice substantial characteristic of periodic waves with which we are quite familiar. We measure frequency in hertz (cycles per second); we can observe frequency on a cathode ray oscilloscope, and 100 Hz has a strong tendency to remain 100 Hz. When our ear perceives it, however, the **pitch** for a soft 100 Hz tone may be quite different than for a loud one. The pitch of a low frequency tone goes down and that of a high frequency tone goes up as intensity is increased. A famous acoustician, Dr. Harvey Fletcher, found that playing pure tones of 168 and 318 Hz at modest level produces a very discordant sound. At a high **intensity**, however, the ear hears them in 150-300 Hz octave relationship as a pleasant sound. For such reasons we cannot equate frequency and pitch, we can only say they are analogous.

The same situation exists between **intensity** and **loudness**, as the relationship between the two is anything but linear. This will be studied in considerable detail in Chapter 2 because of its great importance in high fidelity work.

Similarly, trying to relate measured **waveform** and perceived sound **quality** is complicated by the functioning of the hearing mechanism. As a complex waveform may be described in terms of a fundamental and a train of harmonics of various amplitudes and phases (as we shall see in more detail later), the frequency-pitch interaction would be involved as well as other factors.

In the remainder of this chapter we will deal with the physics of sound in very elementary terms, concentrating on those phenomena of special interest in listening rooms and studios.[2] Chapter 2 treats the ear, how it functions and something of its nonflat, nonlinear characteristics which have to do with our perception and judgment of quality of sounds.

Chapter 3 considers the signals with which we have to deal, both wanted and unwanted. With a reasonable understanding of these three chapters, the problems of high fidelity may be viewed in a new and helpful perspective.

HOW IS SOUND TRANSMITTED?

Imagine, in the laboratory, an electric buzzer suspended within a heavy glass bell jar. As we push the button we hear the buzzer through the glass. As the air is pumped out of the bell jar, however, the sound becomes fainter and fainter until it is no longer audible. The sound-conducting medium, air, has been removed from between the source and the ear. Sound must have a medium or it cannot be transmitted from one point to another. Air is a very common agent for the conduction of sound although other gases as well as solids and liquids conduct sound also. As outer space is an almost perfect vacuum, no sound can be transmitted except in the tiny island of air (oxygen) within a spaceship or a spacesuit.

In Fig. 1-1 we have a weight W suspended from a spring. If the weight is pulled down to the −5 mark and released, the spring will pull it back toward zero. However, the weight will not stop at zero; its inertia will carry it on past zero almost to +5. The weight will continue to oscillate or vibrate at an amplitude that will slowly decrease due to frictional losses within the metal of the spring, in the air, etc.

In the arrangement of Fig. 1-1, vibration or oscillation is possible because of two properties, the **elasticity** of the spring and the **inertia** of the weight. Elasticity and inertia are two things all media must possess to be capable of conducting sound. If an air particle is displaced from its original position, elastic forces of the air tend to restore it to its original position. Because of the inertia of the particle, it overshoots the original position, bringing into play elastic forces in the opposite direction, and so on.

Sound is readily conducted in gases, liquids, or solids such as air, water, steel, concrete, etc., which are all elastic media. As a child, perhaps you have heard two sounds of a rock hitting a railroad rail in the distance, one sound coming through the air and one through the rail. The one through the rail arrived first because the speed of sound in the dense steel is greater

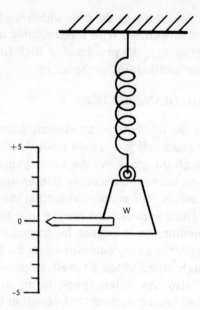

Fig. 1-1. A weight on a spring vibrates at its natural frequency because of the elasticity of the spring and the inertia of the weight.

than that in tenuous air. Sound also travels well in the sea at a speed about five times that in air. The SOFAR system has given pinpoint locations at sea by sounds traveling over a thousand miles through the water. The cacophony of sounds in the sea has been revealed in recent years to be the purposeful, communicating sounds made by whales, porpoises, seals, fish of many kinds, and other creatures.

The Dance of the Particles

Waves created by the wind travel across a field of grain yet the individual stalks remain firmly rooted as the wave travels by. Neither do particles of air propagating a sound wave move far from their undisplaced positions. The disturbance travels on but the particles do not. The particles of the medium do their little dance close to home.

There are three "dances of the particles." If a stone is dropped onto a calm water surface, concentric waves travel out from the point of impact, the water particles describing circular orbits (for deep water at least) as in Fig. 1-2A.

Another kind of wave motion is illustrated by a violin string (Fig. 1-2B). Here the tiny elements of the string move transversely or at right angles to the direction of travel of the wave along the string.

We are primarily interested in sound traveling in a medium such as air. In this case (Fig. 1-2C), the particle vibrates back and forth in the direction the sound wave is traveling. These are called **longitudinal waves**.

How a Sound Wave Moves

Now, how are air particles jiggling back and forth able to carry that beautiful music from the loudspeaker to our ears at the speed of a rifle bullet? In Fig. 1-3 the little dots represent air molecules. This sketch is exaggerated, as there are over a million million million molecules of air in a cubic inch. Fig. 1-3A represents a sound wave traveling from left to right and Fig. 1-3B the same thing an instant later. The molecules crowded together represent areas of **compression** (pressure higher than atmospheric) and the sparse areas represent **rarefactions** (pressure lower than atmospheric). The small arrows tell us that on the average the molecules are moving to the right on the compression crests and to the left in the rarefied troughs between the crests. Any given molecule will move a certain distance to the right and then the same

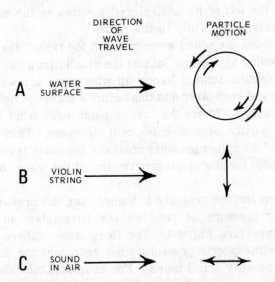

Fig. 1-2. Particles involved in wave motion can dance in circular, transverse, or longitudinal motions.

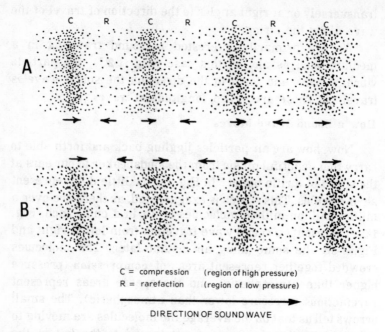

C = compression (region of high pressure)
R = rarefaction (region of low pressure)

DIRECTION OF SOUND WAVE

Fig. 1-3. In A, the sound wave causes the air molecules to be pressed together in some regions and spread out in others. An instant later (B) the wave has moved slightly to the right.

distance to the left of its undisplaced position as the sound wave progresses uniformly to the right.

What makes the sound wave move to the right? The answer is revealed by a closer look at the small arrows of Fig. 1-3. The molecules tend to bunch up where two arrows are pointing toward each other and this occurs a bit to the right of each compression. Where the arrows point away from each other the density of molecules will decrease. Thus the movement of the higher pressure crest and the lower pressure trough account for the small progression of the wave to the right.

At the crests, the pressure is higher than the prevailing atmospheric pressure as read on the barometer; in the troughs, lower (see Fig. 1-4). The fluctuations (above and below the atmospheric pressure) that represent the sound wave may be very small indeed. For example, the faintest sound the human ear can hear (0.0002 microbar) is some 5000 million times smaller than atmospheric pressure.

12

THE SINE WAVE

The weight in Fig. 1-1 moves in what is called simple harmonic motion. If a ballpoint pen is fastened to the pointer and a strip of paper is moved past it at a uniform speed, the resulting trace is a sine wave (Fig. 1-5). The air particles of Fig. 1-4A are set in motion by the tines of a tuning fork which also move in simple harmonic motion and which also create a sine wave pattern of compressions and rarefactions of the air particles, which can be represented by the graph in Fig. 1-4B. The upward loops represent the compressions and the downward troughs the rarefactions, which vary above and below the average static atmospheric pressure prevailing at the moment.

The shape of this graph, the sine wave, is a very special form that has great mathematic, electronic, and acoustic significance. It is the simple form, the elemental building block of which all complex waves are constituted, as we shall see in Chapter 3.

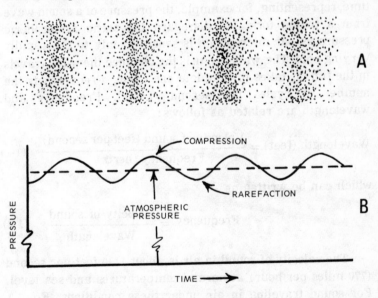

Fig. 1-4. In A, we show an instantaneous view of the compressed and rarefied regions of a sound wave in air. In B, the compressed regions are very slightly above and the rarefied regions very slightly below atmospheric pressure. Pressure variations representing sound waves are thus superimposed on normal barometric pressure.

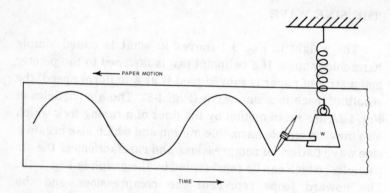

Fig. 1-5. A ballpoint pen fastened to the weight which is vibrating up and down traces a sine wave on a paper strip moving at a uniform speed. This shows the basic relationship between simple harmonic motion and the sine wave.

WAVELENGTH AND FREQUENCY

In Fig. 1-6 a simple sine wave has been plotted against time, representing, for example, the pressure of a sound wave from a tuning fork. We disregard the static atmospheric pressure as our microphone is sensitive only to the rapidly varying part. The **wavelength** is the distance the wave travels in the time it takes to complete one cycle. **Frequency** is the number of cycles per second (hertz). Frequency and wavelength are related as follows:

$$\text{Wavelength (feet)} = \frac{\text{Velocity of sound (feet per second)}}{\text{Frequency (hertz)}} \quad (1\text{-}1)$$

which can be written as:

$$\text{Frequency} = \frac{\text{Velocity of sound}}{\text{Wavelength}} \quad (1\text{-}2)$$

The velocity of sound in air is about 1130 feet per second (770 miles per hour) at normal temperatures and sea level. For sound traveling in air under these conditions, Eq. 1-1 becomes:

$$\text{Wavelength} = \frac{1130}{\text{Frequency}} \quad (1\text{-}3)$$

14

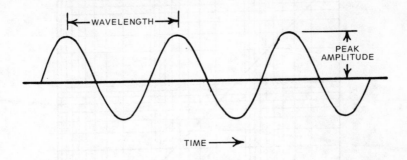

Fig. 1-6. Wavelength is the distance a wave travels in the time it takes to complete one cycle. It may also be expressed as the distance from a point on a periodic wave to the corresponding point on the next cycle of the wave.

We shall have many occasions to use this relationship. Fig. 1-7 is a graph of Eq. 1-3 which will be useful later when we consider listening room and studio dimensions in terms of the wavelength of sound of various frequencies.

INVERSE SQUARE LAW

We have all observed the decrease in the intensity of sound as we move away from the source of sound. For those working with recorders and microphones it is of great practical value to be able to predict the change of sound level to be expected with a given change of distance between sound source and microphone or between loudspeaker and listener. To be able to make such predictions requires some knowledge of how sound spreads out.

There are two main reasons for sound intensity decreasing with distance: (1) losses in the medium and (2) geometrical divergence. For modest distances, losses are so small that we shall neglect them and concentrate on geometrical divergence.

If you spread a given amount of butter over a small piece of bread it will be thicker than if spread over a large slice. So it is with sound. All the sound radiated by the source S in Fig. 1-8 must pass through spheres 1, 2, 3, 4, etc. Taking a small solid angle, the same sound power passing through area A1 also passes through A4. Because A4 is of much greater area than A1, the sound power per square inch at A4 is much less than at

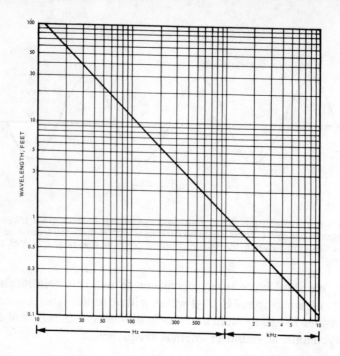

Fig. 1-7. A chart for easy determination of the wavelength in air of sound waves of different frequencies. (Based on a velocity of sound of 1130 ft per second.)

A1. As the area of a sphere is $4\pi r^2$, the sound power per unit area decreases as the square of the distance. This is called the "inverse square law" and accounts only for the geometrical spreading of the sound and is not truly a loss.

Anticipating a later discussion on the decibel, this is the proper place to mention an expression of the inverse square law in very useful form. It can be expressed as "6 dB per distance doubled." For example, if a microphone is 5 feet from an enthusiastic soprano and the VU meter in the control room peaks +6, moving the microphone to 10 feet would bring the reading down **approximately** 6 dB. The word "approximately" is important. The inverse square law holds only for free-field conditions. The effect of sound energy reflected from walls would be to make the change for a doubling of the distance something less than 6 dB.

An awareness of the inverse square law is of distinct help in estimating acoustical situations. For instance, a doubling of the distance from 10 to 20 feet would, for free space, be accompanied by the same sound level decrease, 6 dB, as for a doubling from 100 to 200 feet. This accounts for the great carrying power of sound outdoors.

REFLECTION OF SOUND

Sound is reflected from objects which are large compared to the wavelength of the impinging sound. This book would be a good reflector for 10 kHz sound (wavelength about an inch). At the low end of the audible spectrum, 20 Hz sound (wavelength about 56 feet) would sweep past the book and the person holding it as though they didn't exist and without appreciable "shadows."

Reflected sound follows the same rules as light: the angle of incidence equals the angle of reflection as in Fig. 1-9A. Fig. 1-9B shows how this law of reflection can be put to good use in a parabolic reflector which focuses incoming parallel rays of sound onto a microphone, a good arrangement for the recording of birdsongs or the marching band at a football game.

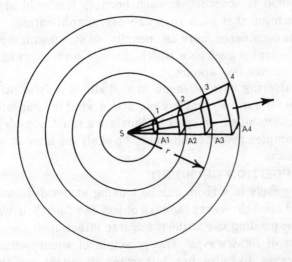

Fig. 1-8. The same sound energy radiated uniformly in all directions from a point source must pass through spheres of increasing area. Therefore, the energy per unit area (and the sound pressure) will vary inversely as the square of the radius of the sphere. Sound falling off according to this "inverse square law" is reduced 6 dB for each doubling of the distance from the source.

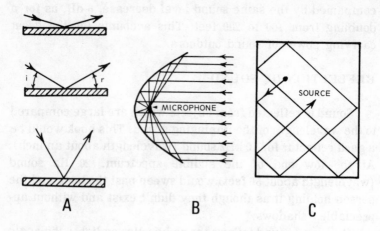

Fig. 1-9. As sound is reflected from a surface, the angle of incidence (i) is equal to the angle of reflection (r). This is true for (A) flat surfaces, (B) a parabolic surface, and (C) the walls of listening rooms and studios.

Fig. 1-9C illustrates how a ray of sound may undergo many reflections in an enclosure such as a studio. How many reflections a single ray would experience depends on how much sound is absorbed at each bounce. It should also be borne in mind that such rays are oversimplifications. They should be considered more as "pencils" of sound with more or less spherical wavefronts which diverge and to which the inverse square law applies.

Considering rays of sound in a studio is helpful up to a point but for a more complete picture of what is happening to our precious music or speech signals we must consider the entire complex reverberant field. We shall see later how this can be done.

SUPERPOSITION OF SOUND

Ten people in a room can be looking at ten different objects and the light waves for each object can find their way to the corresponding eye without adverse effect upon the other nine sets of light waves. The principle of superposition of sound waves likewise has important practical results. It means that the air in a room is capable of carrying many sound waves simultaneously. When an orchestra plays, the waves produced by the clarinet do not disturb those produced by the violin.

If we toss a stone into a quiet pool of water, a series of concentric waves will emanate from the spot where the stone hits the water. If a second stone is tossed in, a second series of waves will be produced. One series will go through the other series with no apparent adverse effect upon either. The principle of superposition (or **interference**) states that the same portion of the medium may transmit simultaneously any number of different series of waves. These proceed independently, each undisturbed by the presence of the others, the displacement of the particles of the medium at any instant being the algebraic sum of the displacements due at that instant to each separate wave system. Let us see what this means in a few specific cases.

In Fig. 1-10A, waves 1 and 2, arriving from different directions, pass the same spot in the medium at the same time and **in phase**. The positive and negative peaks of wave 1 add to the peaks of wave 2 giving a resultant having the same frequency but twice the peak amplitude. In Fig. 1-10B, waves 1 and 2 are of the same frequency and amplitude but wave 2 arrives later in such a way that the two are in what is called **phase opposition**. In this case one wave exactly cancels the other and the resultant is zero. For this particular particle not to move at all is just what is needed to speed both waves 1 and 2 on their respective ways. Of course there is an infinite number of other combinations of amplitude and phase between two such waves.

If two loudspeakers are placed side by side in a room and energized from the same amplifier driven by a sine wave signal of about 500 to 1000 Hz, these constructive and destructive effects are readily observed by walking around in the resultant sound field. For the moment we are interested primarily in demonstrating the actual physical existence of these superposition effects rather than in the details of the sound field, which would be very complex. In performing this experiment with the two loudspeakers several hints are in order: (a) plug one ear to avoid binaural complications, (b) reflections from walls, ceiling, and floor greatly complicate the field, and (c) a handy device for introducing a phase opposition effect as in Fig. 1-10B is to reverse the leads to one of the loudspeakers so that the motion of one cone is out as the other moves in and vice versa. This will shift the location of the maxima and minima in the room.

SOUND AROUND CORNERS

Sound does not always cast sharp "shadows," and the reason is **diffraction**. Normally, sound waves tend to continue traveling in their original direction. When edges of obstacles are encountered, the edge of the barrier becomes a secondary source of sound sending out waves of the same frequency as the original but of lower intensity. These secondary waves spread out into the shadow area. The combination of diffraction and reflection can crowd almost as much sound through a door opened just a crack as through a fully opened door.

ABSORPTION OF SOUND

A reverberation chamber is a room with hard, reflective surfaces. These surfaces are poor absorbers of sound; as a result, the sound decays slowly in such a room. To make a recording studio out of this room would require, among other things, the application of surface materials which would absorb sound efficiently. Absorption of sound is of direct, vital

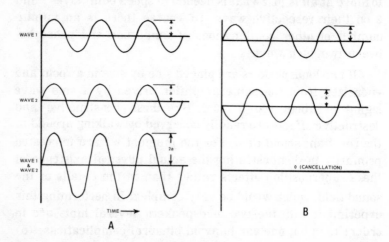

Fig. 1-10. In A, if wave 1 is superimposed on wave 2 and if they are in phase, the amplitude (a) will double. In B, if wave 1 is one half wavelength (180 degrees) out of phase with wave 2, one cancels the other and zero amplitude results.

importance to both technical and artistic personnel engaged in high-fidelity recording and reproducing activities.

When sound strikes a wall, as in Fig. 1-11, some of it is reflected and some of it goes through the wall, emerging from the other side. We must note two things, (a) that the intensity of the sound emerging on the other side is lessened because of the absorption of sound energy by the wall and (b) that in going from air into the wall the sound is refracted back to its original direction as it emerges into the air on the other side of the wall. As these refraction effects do not affect the intensity of the sound, we may neglect them in this discussion.

The sound energy absorbed by the wall is changed to heat energy by the friction encountered by the air particles involved in transmitting the sound. The amount of heat involved is normally extremely small because the amount of energy involved in normal sounds is very small. In fibrous materials such as carpets, drapes, glass wool and common acoustic tile, the sound absorption is high (at least at the higher frequencies) because the sound undergoes many reflections in the fibers and tiny pores, losing energy at each reflection. Many commercial acoustic absorbers are made of such fibrous, porous material.

The fraction of the energy of the incident sound which is absorbed is called the **sound absorption coefficient** of the

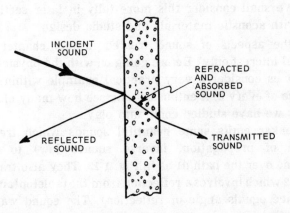

Fig. 1-11. Part of the sound falling on a wall will be reflected. The part penetrating the wall will be bent from the original direction (refracted). That portion of the sound not absorbed within the wall will emerge from the other side traveling in the same direction as the incident sound.

Fig. 1-12. Propagation, inverse square, reflection, superposition of sound are all involved in the simple act of talking into a microphone.

reflecting surface. Hard, massive, nonporous surfaces such as plaster, masonry, glass, wood, concrete, etc., have absorption coefficients of less than 0.05, i.e., absorbing less than 5 percent of the incident energy. Soft, porous materials which permit penetration of the sound waves may have coefficients approaching 1.00 or absorbing close to 100 percent of the incident energy. We shall consider this more fully in later sections dealing with acoustic materials and studio design.

Are the aspects of sound treated in this chapter of theoretical interest only? Before going on with the physics of sound, let us consider a very practical example within the experience of every hi-fi enthusiast and see how many of the principles we have studied come into play.

A speaker emits some beautiful sounds which travel (principle of propagation, inverse square law) to the microphone over the path d1 as in Fig. 1-12. They also travel via path d2 which involves a reflection from the tabletop (angle of incidence equals angle of reflection). The sound waves traveling the two routes combine (superposition) at the diaphragm of the microphone which, in turn, moves in obedience to the pressure changes in the air in contact with it.

In Fig. 1-10B we saw that when waves 1 and 2 were in phase opposition, they canceled each other. Let us take the signal traveling path d1 of Fig. 1-12 as wave 1 and that bouncing off the table as wave 2. The greater length of path d2 means that wave 2 arrives later. At some particular frequency, the added distance over path d2 is just enough to bring wave 2 into phase opposition with wave 1 and the two cancel each other. This means that speech energy near this frequency is lost or seriously reduced. This "destructive interference" occurs at other harmonically related higher frequencies. In between are frequencies at which the sound pressure is increased ("constructive interference" as in Fig. 1-10A). Thus we see that the near presence of the highly reflective surface affects the dulcet tones of the speaker's voice in a most un-hi-fi manner.

This effect may be observed by recording your voice as you hold a microphone a fixed distance from your mouth and then walking toward a hard plaster wall. As you play the test back, a dramatic deterioration of quality is observed as you hear the sound recorded as you approached to within a foot or so of the wall. Moral: keep the microphone away from hard, reflective surfaces or reduce the reflectivity of such surfaces (absorbent table cover, rug under the microphone floor stand, etc.). The other approach used in distant pickups is to mount the microphone very close to the floor so that d1 is essentially equal to d2 and there is minimum interference from the reflected component.[3]

2 | HUMAN HEARING

Every person interested in high fidelity recording and reproduction of sound knows how important the human ear is. Everything revolves around it. "The specs look good, how does it sound?" No matter what the measurements say, the ear is the court of last appeal, the final arbiter. As long as you are buying or building to suit only yourself the problems are minimized—you have only yourself to please. But what hi-fi buff lives only unto himself? Half the fun is bragging about and demonstrating one's latest acquisition to friends. And here we bump into problem 1; the next person may say, "I like my bass a bit fuller than that." Is this just personal preference? Or are his ears different? Or is such a thing as training and experience involved? The answers to these questions are **maybe, yes,** and **undoubtedly!** Let's explore this a bit.

EARS—GOLDEN AND TIN

There is surely such a thing as personal taste in music. Some prefer Beethoven, some Shankar. Assemble all the Beethoven-lovers and one would find that some like the knobs set one way and some another. We would not want to have sterile agreement and uniformity in such things—each person is a unique creation and differences between individuals add zest and interest to our lives.

But there is an important point here that goes beyond esthetic interests and cultural conditioning. Maybe Listener A set the knobs the way he did because his hearing is deficient in the highs! Measurements of hearing acuity have been made upon hundreds of thousands of people and it has been determined that "normal" hearing embraces a considerable spread. It is still very useful to study the characteristics of that "average" ear which, in a sense, describes most of us but which, in another sense, doesn't quite fit any of us.

Listener B may have adjusted the knobs differently because he is a professional who has spent his life not just listening, but listening critically and analytically. Perhaps he rolled off the lows a bit to minimize an irritating room mode, or reduced the volume to control some distortion which Listener A and others didn't even hear. Although the phrase is used sometimes in derision, saying that Listener B has "Golden Ears" would make sense because of his uncanny ability, developed over a decade or two of critical listening, to detect flaws. There may be as much difference between the ability of the man with the golden ears and the average hi-fi enthusiast as there is between the hi-fi enthusiast and the youth walking along the beach absorbed in the raucous output of a transistor radio held to his ear.

Critical and analytical listening can add a tremendous new dimension to one's enjoyment of music, but we should not stop there. If we do, we lose touch with the music itself. This is somewhat like the lighthouse enveloped in pea soup fog. After several days and nights of incessant foghorn blasts, the horn went dead. The lighthouse keeper suddenly looked up at his wife and said, "Sh-h-h, what was that?"

HOW WE HEAR

While the beauty of our external ears may be nothing to brag about, the hearing mechanism of which they are a part is truly a wondrous system.[4,5,6] Capable of hearing, under proper conditions, the infinitely delicate tattoo of air molecules on the eardrum, the ear can handle, without damage, sounds 10 million million times stronger. From the sighing of the leaves in the wind to the roar of a jet engine, the ear can tell one sound from another in an intricate way not yet fully comprehended.

Aural Mechniasm

The external ear helps collect the sound and send it down the canal to the eardrum (Fig. 2-1). The vibrations of this diaphragm of thin skin are transmitted across the air cavity of the middle ear by a linkage of three delicate bones, the **ossicles** (often called the hammer, anvil, and stirrup because

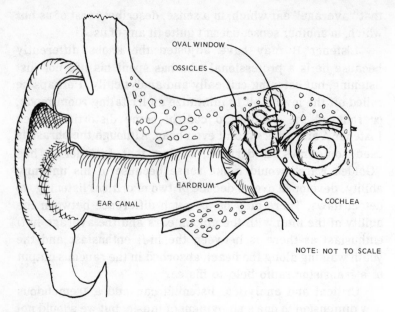

OVAL WINDOW

OSSICLES

EARDRUM

EAR CANAL

COCHLEA

NOTE: NOT TO SCALE

Fig. 2-1. Cross section of the human ear. The airborne sound stimulus causes the eardrum to move. This mechanical movement is transmitted to the liquid within the cochlea by the ossicles. Sensitive hair cells in the cochlea are connected to the brain through the auditory nerve. Amplification of the stimulus takes place both in the ear canal and in the mechanical system of the middle ear (ossicles).

of their shape). Movements of these bones are transmitted to the liquid inside the cochlea through the oval window, a membrane covering the opening to the bony case of the cochlea. Within the cochlea is the sensitive Organ of Corti with its hair cells connected to the auditory nerve which are stimulated in orderly patterns by the sound traveling in the liquid. The entire cochlea is no larger than the tip of one's little finger and there is about a drop of liquid within, surely a marvel of miniaturization.

Because the sounds picked up by the ear are so feeble, they must be handled efficiently in order to reach the cochlea, the true sensing part of the ear, with sufficient magnitude to be detected. The human ear is beautifully designed by the Creator to do just this. First, the ear canal itself is a resonant tube which doubles the pressure acting on the eardrum. Secondly, the lever action of the ossicles provides another increase in force of up to 3 times. Thirdly, the ratio of areas of the relatively large eardrum and the tiny oval window to which the "stirrup" bone is attached provides as much as another 30-fold increase. The combination of all these "force amplifiers" provides a total increase of around 200 which is just what is needed to convey sound waves from light and compressible air to sound waves in the dense and in-compressible liquid of the cochlea. The electronics engineer will recognize this as impedance matching. It is interesting to note that the ossicles of a baby are fully formed and do not grow appreciably as the baby grows. To do so would upset the efficiency of this wonderful listening mechanism.

We have discussed only the mechanical aspects of the ear which are quite well understood. The intricate operation of the cochlea itself is currently under intense investigation and is not yet clearly understood. One thing that seems clear, however, is the fact that the cochlea transforms the mechanical energy of the acoustic signals to coded electrical signals which are sent to the brain. In the process the cochlea acts as some sort of analyzer of sound. The cochlea of Fig. 2-1 is rolled up like the snail's shell from which it gets its name. Let us imagine unrolling it and stretching it out to its full length of about 32 mm. Tones of different frequencies result in points of maximum response at different positions along the cochlea. A high-frequency tone will create a maximum effect in the cochlea near the oval window. A low-frequency tone will create a maximum near the extreme tip of the straightened cochlea. Let us say that a 250 Hz tone (about middle C) gives rise to a maximum at a certain point near the tip. Doubling the frequency to 500 Hz (one octave) moves the maximum about 5 mm toward the oval window. Going up another octave to 1000 Hz moves the maximum approximately another 5 mm. The mathematician would spot this as a **logarithmic relationship** between frequency and position along the cochlea. A similar frequency-position relationship exists in the acoustic cortex of the brain. This suggests that there are valid physiological reasons behind our use of a logarithmic scale for frequency and using the logarithmic unit, the decibel.

It is in the cochlea that sound waves are converted to nerve impulses and passed on to the brain by the auditory nerve fibers. These impulses are in code. The journey of the sound ends in the brain, where it is decoded, only an in-stant after it enters the ear canal.

Loudness

In Chapter 1 it was suggested that there is no one-to-one relationship between sound intensity, a physical quantity, and loudness perceived by the human hearing apparatus. Fig. 2-2 is a refinement of the famous Fletcher-Munson **equal-loudness contours** which show that loudness and sound pressure are related in a complex, contorted way. These contours result from measurements on many people and may be taken as characteristic of "average" hearing in humans. But how are the curves obtained?

Fig. 2-2 displays a family of equal-loudness contours of the human ear responding to pure tones at many loudness levels. The loudness level of 10 phons is arbitrarily made to correspond to a sound pressure level of 10 dB at 1000 Hz (1 kHz). This is the definition of the **phon**. A 100 Hz tone must have a sound pressure level 20 dB higher, or 30 dB, to be judged to have the same loudness as the 1 kHz tone. The 10-phon loudness contour is located by plotting the sound pressure level required experimentally to make the tones of various frequencies as loud as the 1 kHz tone. The 20-phon loudness contour and the others are built up in the same way.

This discussion is certainly not the last word on the subject of loudness. People agree quite well when asked to estimate when one sound is twice as loud or half as loud as another. But this kind of subjective judgment of loudness is not well described by the phons of Fig. 2-2, which were defined quite arbitrarily. Researches have come up with another unit, the **sone**, which comes closer to measuring what we perceive but a consideration of the sone is quite beyond the scope of this book. It should be remembered that Figs. 2-2, 2-3, 2-4 are derived from comparisons of pure tones and that a comparison of the loudness of two complex sounds is quite a different problem and a much more complicated one.

Limits of Hearing

No sounds are audible to the average human ear if they have sound pressure levels less than the "minimum audible" curve. This represents the threshold of hearing. If any particular tone is gradually increased in intensity, a point is reached at which the sensation of sound gives way first to a tickling sensation and then to pain. This occurs at about the 130 phon curve of Fig. 2-2.

The "auditory area" of Fig. 2-3 is bounded below by the **threshold of hearing** (or "minimum audible") and above by the **threshold of feeling** or pain. All the tones of every frequency and intensity to which the human ear is capable of responding are contained within this area. We shall consider this auditory area further in Chapter 3 with respect to speech and music.

Frequency Response of the Ear

We can strain hard for flat response of amplifiers, microphones, and loudspeakers, but the stark fact is that the frequency response of the ear is not flat. In fact the ear's frequency response changes with loudness of the sound, a condition that would wreak havoc in the electronic area! A knowledge of this characteristic of the ear is of great importance to all who are called upon to judge sound quality.

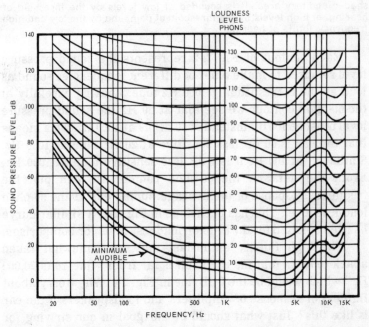

Fig. 2-2. Equal-loudness contours of the average human ear, commonly called the Fletcher-Munson curves. The 70-phon contour shows the levels of sound required at each frequency to make it sound as loud as the 70 dB tone at 1000 Hz. These curves show the lack of flat response in the ear.

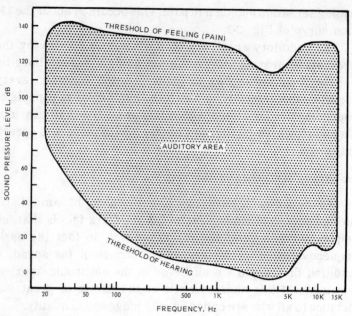

Fig. 2-3. All the sounds of life which we can perceive must fall within the shaded auditory area. It is bounded at low levels by the threshold of hearing, at high levels by the threshold of pain, and by the low and high frequency limits of hearing.

Each contour of Fig. 2-2 represents the sound pressure level required to make tones of different frequencies sound as loud as the corresponding 1000 Hz tone. Looking carefully at the 60-phon contour we see that at 20 Hz the sound pressure level must be 100 dB instead of 60 dB as at 1000 Hz. This means that the ear is 40 dB less sensitive at 20 Hz than at 1000 Hz. Similarly, at 100 Hz it is 8 dB less sensitive than at 1000 Hz. From the data on the 60-phon contour we can plot the frequency response of the average human ear at the 60-phon loudness level. This has been done in Fig. 2-4. A similar curve has been plotted for the 80-phon loudness level for comparison.

The high fidelity enthusiast may be somewhat taken aback to learn that the ear itself is far from "flat, 20 to 20,000 Hz" which would be the 0 dB line in Fig. 2-4. Why worry about flatness of response of amplifiers and loudspeakers if the ear is like this? Just what should be our goal in our striving for perfection in reproduction and transmission of sound?

Let us go to the concert hall, sit in the best seat, and listen to a symphony orchestra. We hear the music and enjoy it very much, not particularly conscious of problems of dynamic range, frequency response, loudness level, etc. Perhaps our goal should be to reproduce in our home the music as heard in the concert hall as perfectly as possible.

Let us say that the high fidelity enthusiast adjusts the volume control on his amplifier so that the level of the recorded symphony music is pleasing as a background to conversation (assumed to be about 60 phons). As the passage was played at something like an 80-phon loudness level in the concert hall, something needs to be done to give the bass and treble of the music the proper balance at the lower-than-concert-hall level. Our enthusiast would find it necessary to increase both bass and treble for good balance. The "loudness control" found on most amplifiers adjusts electrical networks to compensate for the change in frequency response of the ear for different loudness levels.[7] This discriminating hi-fi buff, however, may not be very happy if this loudness compensation is locked to the position of the volume control which is not necessarily closely related to the perceived loudness of the sound. How can one simple loudness control care for so many variables? So perhaps we have not reached the ultimate in the design of a loudness control. The volume control setting is

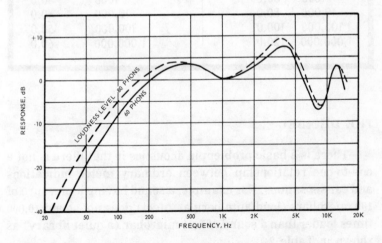

Fig. 2-4. The frequency response of the human ear is not flat. Further, the ear's response varies with loudness of the sound, being flatter for very loud sounds than for soft sounds. The loudness control on hi-fi amplifiers is intended to compensate somewhat for this characteristic of the ear.

affected by the output of the phono cartridge, the sensitivity of the power amplifiers, the efficiency of the loudspeakers, as well as the vagaries of human hearing.

Table 2-1.

ELECTRICAL Current or Voltage Ratio		ELECTRICAL Power Ratio	
ACOUSTICAL Pressure Ratio	dB	ACOUSTICAL Sound Power or Intensity Ratio	dB
1	0.0	1	0.0
2	6.0	2	3.0
3	9.5	3	4.8
4	12.0	4	6.0
5	14.0	5	7.0
6	15.6	6	7.8
7	16.9	7	8.5
8	18.1	8	9.0
9	19.1	9	9.5
10	20.0	10	10.0
100	40.0	100	20.0
1000	60.0	1000	30.0
10,000	80.0	10,000	40.0
100,000	100.0	100,000	50.0
1,000,000	120.0	1,000,000	60.0

THE DECIBEL

There is a basic problem in acoustics in that there is not a one-to-one relationship between ordinary meter indications and ear sensations. For example, a sound having a pressure of 100 microbars (loud auto horn at 3 feet) does not sound 10,000 times louder than a sound of 0.01 microbar (a quiet library) as shown in Table 2-1.

Another problem arises from the tremendous range of the human ear. The weakest sound the average person can hear has a pressure of about 0.0002 microbar. The loudest sound which can be heard without being painful is about 1000 microbars. This range of pressures is over a million to one. Working with ordinary numbers spanning such a range is awkward and inconvenient.

Thus both convenience and the way the ear works require a better method of handling acoustic data. The decibel concept has answered both needs very effectively. It is not necessary to understand all the mathematics behind it to use the decibel, but for the more technical reader the decibel is defined as "ten times the logarithm to the base ten of the ratio between two power quantities." As sound power is related to the square of the sound pressure, a convenient scale for sound measurements is:

$$\text{Sound pressure level} = 10 \log \left(\frac{p^2}{p_0{}^2} \right)$$

$$= 20 \log \left(\frac{p}{p_0} \right) \text{ in decibels}$$

where p is the sound pressure being measured and p_0 is an agreed upon reference pressure, 0.0002 microbar. The word "level" means that the pressure being examined is expressed in terms of dB above some standard reference level.

Since 1948 there has been agreement among electrical enginners on 1 milliwatt (1 mW = 0.001 watt) as the reference level. This has given rise to the designation dBm. Thus a calibrated VU meter reading −10 dBm would indicate an electrical level of 10 dB below 1 milliwatt. The 0.0002-microbar reference level is close to that acoustic pressure which can just barely be perceived by the average person. Thus a sound pressure level of 80 dB would indicate a sound pressure 80 dB above this threshold.

In the light of the above definition of the decibel and common examples, we are forced to consider ratios when dealing with decibels. If the input of an amplifier is 0.01 volt and the output is 10 volts, the voltage gain of the amplifier would be 10 divided by 0.01, or 1000. From Table 2-1, we see that this corresponds to 60 dB. If the pressure of sound in a studio measured with a calibrated system were found to be 0.8 microbar, the ratio with the reference level would be 0.8

Table 2-2.

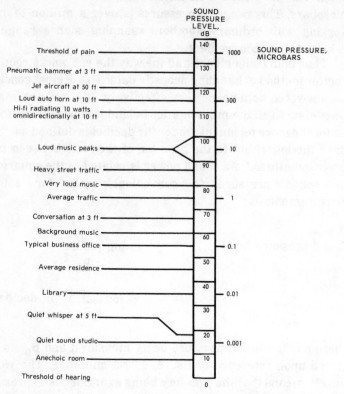

SOUND PRESSURE LEVEL, dB

Threshold of pain	140	1000 SOUND PRESSURE, MICROBARS
Pneumatic hammer at 3 ft	130	
Jet aircraft at 50 ft	120	100
Loud auto horn at 10 ft		
Hi-fi radiating 10 watts omnidirectionally at 10 ft	110	
Loud music peaks	100	10
Heavy street traffic	90	
Very loud music		
Average traffic	80	1
Conversation at 3 ft	70	
Background music		
Typical business office	60	0.1
Average residence	50	
Library	40	0.01
Quiet whisper at 5 ft	30	
Quiet sound studio	20	0.001
Anechoic room	10	
Threshold of hearing	0	

0.0002 MICROBAR (REFERENCE)

divided by 0.0002, or 4000. Referring to Table 2-1, again, we see that a ratio of 1000 would give 60 dB and a ratio of 4 would give 12 dB; therefore, the sound pressure **level** would be 72 dB.

To give us a "feel" for sound levels, Table 2-2 lists some familiar sounds with the corresponding sound pressures and sound pressure levels. A million-to-one range in sound pressure is conveniently reduced to a 120 dB range in sound pressure levels.

THE DISCRIMINATION ABILITY OF THE EAR

The faders on mixing consoles are commonly built with 2 dB steps. If these were 5 dB steps, movement of the fader would give very noticeable step-wise increments in sound which would be very disturbing. If they were 0.5 dB steps the fader would be too expensive. The 2 dB steps were selected

34

because steps of this magnitude are generally just barely detectable by the human ear. This is only approximately true. Detecting differences in intensities varies somewhat with frequency and also with sound level.

At 1000 Hz, for very low levels, a 3 dB change is the least detectable by the ear but at high levels the ear can detect a 0.25 dB change. For a very low level 35 Hz tone, 9 dB is the least detectable. However, for the important mid-frequency range and for commonly used sound levels, the minimum detectable change in level which the ear can detect is about 2 or 3 dB. Fussing around trying to make level adjustments less than this is an exercise in futility.

Those who work in sound are interested in the ear's power of discrimination in frequency or pitch as well as intensity. Experimenters have found that with moderately loud sounds below 1000 Hz we can detect a change in frequency of about 3 Hz. Above 1000 Hz the minimum detectable change in frequency turns out to be a constant percentage of the frequency and amounts to about one semitone of the musical scale.

Researchers tell us that there are about 280 discernible steps in intensity and some 1400 discernible steps in pitch which can be detected by the human ear. As changes in intensity and pitch are the very stuff of communication, it would be interesting to know how many combinations are possible. Offhand it might seem that there would be 280 times 1400 or 392,000 combinations detectable by the ear. This is overly optimistic because the tests were conducted by comparing two simple, single-frequency sounds in rapid succession and bears little resemblance to the complexities of commonly used sounds. Actually, more realistic experiments show that the ear can detect only about 7 degrees of loudness and 7 degrees of pitch or only 49 pitch-loudness combinations. This is not too far from the number of **phonemes** (the smallest unit in a language that distinguishes one utterance from another) which can be detected in a language.

THE EAR AS AN ANALYZER

Let us perform an experiment. Listening to a good recording of a symphony orchestra, concentrate your attention on the first violins. Now focus attention on the clarinets, then the drums. Next listen to a male quartet and single out the first tenor, the baritone, the bass. This is a very remarkable power of the human ear-brain combination. In the ear canal all these sounds are mixed together; how does the ear succeed in separating them? The sea surface may be disturbed by many wave systems, one due to the local wind, one from a distant storm, and several wakes from passing vessels. The eye cannot separate these, but this is essentially what the ear is constantly doing with complex sound waves. In fact, by rigorous training a keen observer can listen to the sound of a violin and pick out the various harmonics apart from the fundamental!

BINAURAL HEARING

Stereophonic records and sound systems are a relatively new development. Stereo hearing has been around at least as long as man. Both are concerned with the **localization** of the source of a sound. In early times some people thought that having two ears was like having two lungs or two kidneys, if something went wrong with one the other could still function. Lord Rayleigh laid that idea to rest by a simple experiment on the lawn of Cambridge University. A circle of assistants spoke or struck tuning forks and Lord Rayleigh in the center with his eyes closed pointed to the source of sound with great accuracy, thus confirming the fact that two ears function together in **binaural localization**.

It turns out that actually two factors are involved, the difference in intensity and the difference in time of arrival (phase) of the sound falling on the two ears. In Fig. 2-5 the ear nearest the source receives a greater intensity than the far ear because of the "sound shadow" (diffraction) caused by the head. Because of difference of distance to the source, the far ear receives sound somewhat later than the near ear. Below 1000 Hz the phase effect dominates, while above 1000 Hz the intensity effect dominates. There is one localization blind spot. A listener cannot tell whether sounds are coming from directly in front or from directly behind because the intensity of sound arriving at each ear is the same and in the same phase.

In both World Wars aircraft detection and location systems utilized the binaural effect. A pair of horns picked up the sound which was conducted to the ears by stethoscope

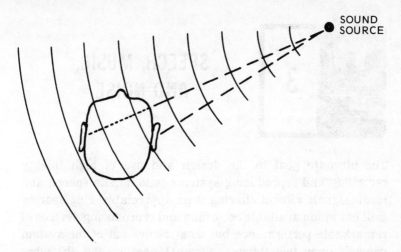

SOUND
SOURCE

Fig. 2-5. Our binaural directional sense is dependent on the difference in intensity and phase of the sound falling on the two ears.

tubes, left horn to left ear, right horn to right ear. By exaggerating the separation of the horns, the phase effect could be heightened and sharper bearings could be obtained. The American "S-Boat" submarines in World War I had a similar system for underwater detection. In place of horns, rubber spheres, separated several feet and in contact with the water, were connected to the stethoscope tubes.

It is interesting that two ears hear more than one. A sound striking two ears seems louder than a sound in one ear (though not twice as loud). Although opening and closing one eye does not change the apparent brightness of the scene, the outputs of the two ears do add.

When we are in a noisy room or one plagued with too much reverberation, we are less irritated by the defects than when we hear such sounds over a "one-eared" (monaural) channel such as radio or television. In the same room our binaural sense of direction can be brought into play, concentrating on the speaker and rejecting noise, echoes, and reverberation. This is one good reason radio and television studios as well as recording studios must have good acoustics and low noise levels.

3

SPEECH, MUSIC, AND NOISE

The ultimate goal in the design and use of high fidelity recording and reproducing systems is to handle speech and music signals without altering them appreciably. Engineering skill has made available recording and reproducing devices of remarkable performance but what comes out of the system depends upon two things, (a) what goes in and (b) what happens to it as it passes through the system. In this chapter we deal with the complexities of the signals applied to the input and something of the types of **distortion** which could occur within, but we shall let the electronics engineer worry about the rest of it. Speech, music (except synthetic), and some noise originates in acoustic space with which this book is chiefly concerned.

SPEECH

The energy involved in speech is amazingly low. A normal conversational voice registering a level of 65 dB at a distance of 3 feet from the person speaking has an average power of 50-millionths of a watt. That is, it would take the simultaneous power of a million such conversationalists to light a 50-watt electric lamp! However, lighting lamps with hot air is not as interesting to the hi-fi specialist as is the distribution of speech power with frequency.

As much as 95 percent of speech power is concentrated below 1000 Hz and yet the other 5 percent is of extreme importance in understanding speech. As we are confronted with tone controls and equalizer knobs in a speech circuit, it is well to remember that, in general, most of the power of speech is in the low frequencies. But good intelligibility is dependent upon the presence of that very important, small portion of power in the highs.

What is required for high fidelity reproduction of speech? Certainly a good signal-to-noise ratio and freedom from distortion are assumed. Beyond this, near-perfect understandability and voice recognition might be listed. All of these required features are the very fabric of high fidelity techniques. However, good high frequency response can create problems. For instance, some voices have an overabundance of sibilants, sometimes caused by dentures, sometimes a natural result of a particular mouth shape, tongue, and set of teeth. These high frequency sibilants may require some "de-essing" or rolling off of the highs to bring them under control. Because of the greater directivity of the sibilants, having a speaker direct his voice to one side of the microphone may achieve similar results.

In Fig. 3-1 we see that speech occupies only a small fraction of the total auditory area. Studying this speech area we see that a frequency range from about 100 to 8000 Hz is sufficient for reproduction of speech with perfect fidelity.

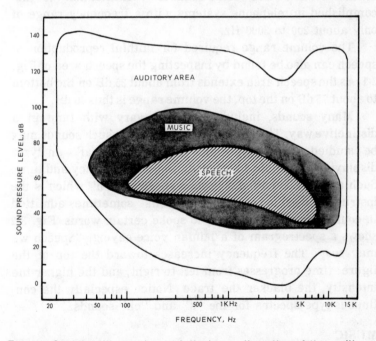

Fig. 3-1. Speech utilizes only a relatively small portion of the auditory area, music considerably more. These two areas specify the frequency range (horizontally) and the dynamic range (vertically) required for speech and music.

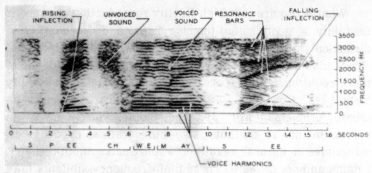

Fig. 3-2. A sound spectrograph analysis from Bell Telephone Laboratories reveals how complicated speech sounds really are. Intensity is indicated by the darkness of the tracing.

What happens if our system frequency range is narrower than this? It has been found that good understandability of speech (high "articulation") can be obtained with a surprisingly narrow frequency range, although, of course, at the expense of quality. The telephone is a good example of this. Practical communication and fair speaker identification is accomplished in telephone systems with a frequency range of only about 200 to 3000 Hz.

The volume range required for faithful reproduction of speech can also be found by inspecting the speech area of Fig. 3-1. As the speech area extends from about 35 dB on the bottom to about 75 dB on the top, the volume range is thus 40 dB.

Many sounds, including speech, vary with time in a distinctive way. The distribution of energy in such sounds may be studied with a special "three-dimensional" analyzer displaying variations in intensity in both frequency and time. Such a device is called a "sound spectrograph" which is the instrument used to record "voiceprints" sometimes admitted in courtrooms to establish who spoke certain words. Fig. 3-2 shows a spectrogram of a human voice saying, "Speech we may see." The frequency increases toward the top of the figure; time progresses from left to right, and the higher the intensity, the blacker the trace. Notice especially the continuous-type spectra for the "s" and "ch" sounds.

MUSIC

The music area of Fig. 3-1 is considerably greater than the speech area. In other words, as every audiophile knows, a greater frequency range and a greater volume range are required for the faithful reproduction of music than for speech. And yet, a musical selection which demands all of the music area comes far from exercising the full capability of the human ear. From Fig. 3-1 we can see that the frequency range of music extends from 40 Hz to 14,000 Hz and the volume range is about 70 dB.

The frequency range of different musical instruments is shown diagramatically in Fig. 3-3. Why is it that the cello sounds so very different from the trombone even though they have about the same frequency range? The answer lies in the **distribution** of sound energy throughout this range. All of these instruments produce line spectra of the type shown in Fig. 3-4C and the relative intensity and phase of the harmonics are responsible for the distinctiveness (timbre) of each.

DYNAMIC RANGE

The dynamic range of a device is the usable intensity range bounded at the lower extreme by noise and at the upper extreme by distortion. In Figs. 2-2 and 2-3 we have seen that the ear can effectively handle a tremendous range of sound

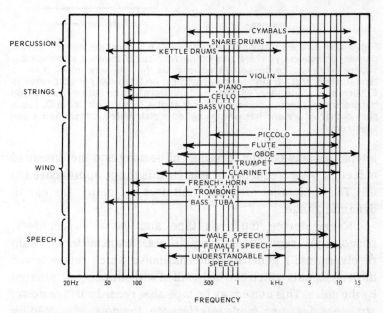

Fig. 3-3. The frequency range of speech and common musical instruments. Two instruments having the same frequency range may sound quite different because of difference in harmonic structure.

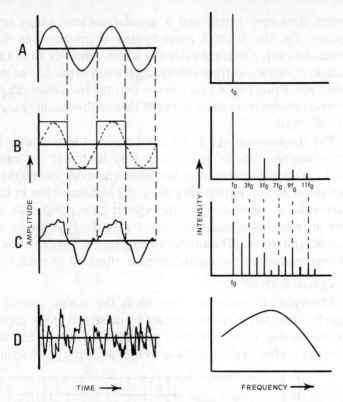

TIME ⟶   FREQUENCY ⟶

Fig. 3-4. In A, the energy of a pure sine wave is concentrated at the fundamental frequency; there are no harmonics. B shows a symmetrical square wave having the same frequency as the sine wave of A; it has a fundamental and a train of odd harmonics. An irregular periodic wave in C, such as a violin tone, will have a fundamental and both even and odd harmonics. A random noise, such as that of a jet aircraft, as in D, has a periodic structure and has what is called a **distributed** rather than a **line** spectrum.

intensities between the threshold of hearing and the threshold of feeling. At a frequency of 1000 Hz this range approaches 130 dB. The equipment man has built does not match the ear in dynamic range.

Every device (such as tape and cassette recorders, phonograph reproducers, amplifiers, transmitters, radio receivers, etc.) generates internal noise which sets a lower limit on signals which can be handled without being dominated by the noise. This noise may be tape hiss, record surface noise, atmospherics, man-made interference, the noise produced by the thermal agitation of electrons flowing in conductors, power circuit hum, or a combination of these. It is interesting that the thermal noise level of air molecules acting on the diaphragm of a very sensitive microphone is of the same order of magnitude as the thermal noise level in electrical conductors. Thus the smallest signal which can be handled by a communication system is fixed by the ever-present noise.

Every recording and reproducing device is also limited in how strong a signal it can handle. This upper limit of signal strength is set by distortion. As the signal level is increased, a point is eventually reached at which the acceptable distortion limit is exceeded. This distortion might be caused by microphone overload, tape saturation, or by signals that exceed the capabilities of a transformer or a transistor somewhere in the chain.

Although the dynamic range of commonly available equipment is sufficient for speech, some form of compression is necessary to reduce the 70 dB range of music to fit within the limitations of practical channels. This can be done manually by raising the soft passages and reducing the loud ones, or automatically by special electronic circuits. Composers, conductors, and other musically knowledgeable persons cringe at what can happen to good music in this necessary process whether it is accomplished manually or electronically.

LINE SPECTRA

The banker must know money, the doctor must know human physiology, and the worker in sound, whether he be in the technical or artistic side of the business, should know the characteristics of the sound with which he deals if he is to do his work well. The time-worn approach, "It is all right if it sounds right to me," might have some validity for presentations to audiences in person, but if technical channels are used, their effect on the signal must always be considered. For this reason we shall study a few types of signals and their characteristics.

The 10 octaves of sound from about 20 Hz to 20,000 Hz is the keyboard upon which the sounds of human life are played. One sound **sounds different** from another sound because of different time patterns and because of different distributions of energy throughout this frequency range, or spectrum. Let us first direct our attention toward so-called "steady-state" waves which repeat themselves cycle after cycle (periodic waves).

The form of the simplest wave, as we have seen, is the "pure" sine wave. All of the energy of this type of wave is concentrated at a single frequency, f_0, as illustrated in the right hand graph of Fig. 3-4A. There are no harmonics. This is commonly referred to as a **line spectrum** whether the energy is concentrated in one or more lines.

A square wave of the same frequency as the sine wave is shown in Fig. 3-4B. In this case there is also an f_0 spectrum line which represents the energy of a pure sine wave of fundamental frequency f_0 shown inscribed with broken line within the square wave. Because the square wave is something other than a pure sine wave, harmonics are present, in this case a series of odd harmonics, $3f$, $5f_0$, $7f_0$...etc., represented by lines of decreasing amplitude.*In other words, if the spectrum of the square wave were searched out with a narrow filter moved up and down in frequency, the spectral lines of energy of Fig. 3-4B would be found and their relative intensities could be measured. Conversely, the square wave can be built up by combining pure sine waves of frequencies f_0, $3f_0$, $5f_0$,$7f_0$,...etc., with proper amplitudes and with proper phase relationships. In this sense the sine wave is considered a basic building block from which complex periodic waves of any shape can be built. Stated another way, comples periodic waves can be broken down into their sine wave components.

In Fig. 3-4C we see the wave shape of a certain note produced by a violin. Its spectrum consists of both odd and even harmonics and the relative intensities of these harmonics are determined by the construction and quality of the instrument. The greater the harmonic content, the richer the tone, or timbre.

---

* There is often confusion between the musician's term "overtones" and the physicist's "harmonics." Overtones occur at successively higher octaves. A harmonic (sometimes called a "partial") is any whole number multiple of the fundamental frequency. With a fundamental of 100 Hz, the octaves would be 200, 400, 800...etc. Hz while the harmonics would occur at 200, 300, 400, 500, 600 etc. Hz.

44

CONTINUOUS SPECTRA

The periodic waves of Figs. 3-4B and C have energy concentrated at frequencies related harmonically to the fundamental frequency. The irregular, nonperiodic wave of Fig. 3-4D is that of the noise of a jet aircraft engine. In this case the energy is continuous through a wide band of frequencies and is called a "distributed" spectrum. It is made up of an infinite number of sine components continually shifting in amplitude and phase in a random way.

There are other sounds difficult to classify. The clanging sound of a bell has many high frequency components which are not integral multiples (harmonics) of the lowest frequency. There may be certain components close enough to harmonic relationship, however, to produce a sound with a definite pitch.

We frequently encounter combinations of continuous and line spectra. A submarine under way produces a potpourri of noises including a throbbing, distributed noise produced by the propeller. Superimposed on the distributed spectrum may be one or more spectral lines associated with the whine of auxiliary machines.

PHASE RELATIONSHIPS

Let us examine the effect of phase relationships which have been mentioned previously. In Fig. 3-5A a fundamental (f_0), a second harmonic $(2f_0)$, and a third harmonic $(3f_0)$, of the indicated amplitudes are combined to form the complex periodic wave at the top of the figure. The reader can demonstrate this to his own satisfaction by combining the instantaneous amplitudes of f_0, $2f_0$, and $3f_0$ at a given time to find the instantaneous amplitude of the resultant wave. For example, at time t1 all the waves are passing through zero, which gives a zero resultant. At time t2, $3f_0$ is equal to zero, $2f_0$ is a positive amplitude, and f_0 is a somewhat larger negative amplitude; the result is a slightly negative value for the resultant complex wave. For time t3 the fundamental and both harmonics are positive, yielding a high positive excursion in the resultant wave.

We have an identical situation in Fig. 3-5B except that both $2f_0$ and $3f_0$ have been shifted to the left a quarter of their respective wavelengths. The second and third harmonics now

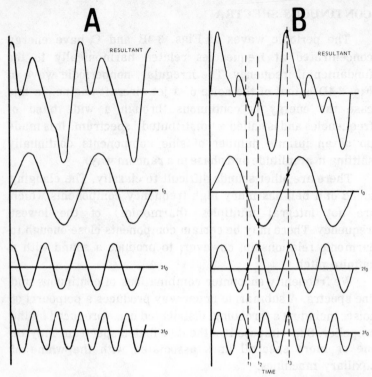

Fig. 3-5. The relative phase of components of a complex periodic wave can have considerable effect on the appearance of the resultant wave on a cathode ray oscilloscope. In A, a fundamental and two harmonics add to give the resultant pictured. In B, everything is the same except that the $2f_0$ and $3f_0$ components have been shifted in phase, affecting the shape of the resultant wave. The ear, being relatively insensitive to phase shifts, might find it difficult to detect a difference between the resultants of A and B.

have a different time relationship to the fundamental but all frequencies and amplitudes are the same as before. Adding ordinates of the three waves, f_0, $2f_0$, and $3f_0$, we get a resultant which has quite a different waveshape. The difference must be attributed to the only thing that has been changed: the time relationship. These relative time shifts are called "phase shifts." Phase shifts may be unintentionally introduced at many places in the communication chain, e.g., in electronic devices, in microphones, or in loudspeakers. The two resultant waves of Figs. 3-5A and B may sound quite similar to the ear in spite of their pronounced different shape because the human ear is relatively insensitive to phase differences.

Much can be learned about the nature of a complex sound signal by breaking it down into its constituent sine components. This is commonly done in the laboratory with a "frequency analyzer," a specialized electronic measuring instrument capable of sorting out the harmonics of a complex periodic wave.

NOISE—THE BAD KIND

Noise is always present in all communication channels, whether it be a cozy conversation over a cup of coffee, watching an international television program arriving via satellite, or reading the morning newspaper. In this general sense "noise" may be taken to be any disturbance tending to interfere with the communication that is going on. This is "bad" noise which we have already considered in our discussion on dynamic range.

NOISE—THE GOOD KIND

A good kind of noise? Defining noise as unwanted sound fits the system noise described above, but noise is becoming an increasingly important tool for measurements in acoustics as we shall see in a later chapter. The noise isn't necessarily different from the bad noise above, but it is put to a specific use.

In acoustical measurements the use of pure tones might be very difficult to handle while a narrow band of noise centered on that same frequency would make satisfactory measurements possible. For example, if a studio is filled with a pure tone signal of 1000 Hz from a loudspeaker, a microphone picking up this sound will have an output that varies greatly from position to position due to room resonances. If, however, a band of noise one octave wide centered at 1000 Hz were radiated from the same loudspeaker, the level from position to position would tend to be more uniform, yet the measurement would contain information on what is happening in the region of 1000 Hz. Such measuring techniques make sense as we are usually interested in how a studio or listening room reacts to the very complex sounds we are recording or reproducing, rather than to steady, pure tones.

Random Noise

Random noise[8] is generated in any electrical circuit and minimizing it often becomes a very difficult problem. Heavy

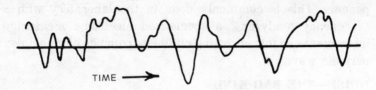

TIME ⟶

Fig. 3-6. A cathode ray oscilloscope shows a section of random noise spread out in time. No periodic repetition is present. In other words the fluctuations are random.[8]

ions falling back on the cathode of a thermionic vacuum tube produce noise of relatively high amplitude and wide spectrum and the introduction of some gas molecules into the evacuated space will produce even more noise. Today a random noise generator is made with a silicon diode or other solid-state device followed by an amplifier, voltmeter, and attenuator.[9]

Random noise would appear on a cathode ray oscilloscope as in Fig. 3-6. Noise is said to be purely random in character if it has a "normal" or "gaussian" distribution of amplitudes. This simply means that if we sampled the instantaneous voltage, say, at a thousand equally spaced times, some readings would be positive, some negative, some greater, some smaller, and a plot of these samples would approach the familiar gaussian distribution curve of Fig. 3-7.

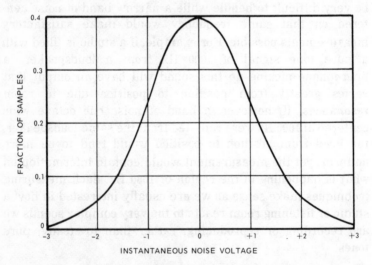

INSTANTANEOUS NOISE VOLTAGE

Fig. 3-7. The proof of randomness of random noise is in sampling the instantaneous voltage, say, at 1000 points equally spaced in time and plotting the results. The familiar bell-shaped gaussian distribution curve results if the noise is truly random.[8]

White and Pink Noise

The energy of such random noise is spread more or less uniformly over a wide frequency range. Actually, the shape of the noise spectrum is determined somewhat by the type of analyzer used in determining the distribution of energy. If the noise is analyzed with a filter which has a fixed bandwidth which could be tuned to any frequency in the audible spectrum, we would find that the analyzer output would be essentially the same no matter to what frequency it is tuned. In other words, the noise spectrum would be as shown in Fig. 3-8A. If the random noise signal is analyzed by an analyzer which has a filter whose passband width is a given percentage of the frequency to which it is tuned, the analyzer output would increase with frequency. In Fig. 3-8B the same noise has been analyzed with a tunable filter one-third octave wide. At 100 Hz the bandwidth is only 23 Hz but at 10 kHz the bandwidth is 2300 Hz. Obviously there would be much greater noise energy in a one-third octave band centered at 100 Hz than one centered at 10 Hz. This means that analyzing the same noise signal with a one-third octave analyzer would yield a spectrum which rises with frequency at 3 dB per octave.

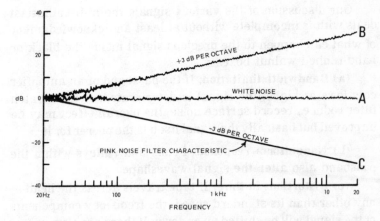

Fig. 3-8. If the spectrum of random noise is measured by an analyzer of fixed bandwidth (a very common type), the spectrum is constant with frequency as in A. If measured with an analyzer whose passband width is a given percentage of the frequency to which it is tuned, the spectrum will slope upward at 3 dB per octave as in B. By putting white noise through a pink noise filter (C) we obtain a signal that yields a flat output which is useful when measurements are made with, for example, one-third octave filters.[8]

As the one-third octave system is so convenient and so widely used in acoustic measurements, it is customary to process the noise of Fig. 3-8 by passing through a "pink noise filter" having a minus 3 dB per octave slope as in Fig. 3-8C which brings graph B down to the "white noise" characteristic of graph A and thus well adapted to many acoustic measurements. Or, this filter can convert white noise to pink noise for convenience in acoustic measurements involving, for instance, one-third octave filters.

The application of color terms to noise may sound quaint, but it comes about quite naturally. White light is a mixture of all colors. If white light is analyzed with optical filters, various colors will result. The low frequency (long wavelength) end of the visible spectrum is red, the high frequency end is violet. Thus "white" noise has a continuous distribution of energy like white light. Noise energy distributed throughout the audible spectrum, but having the low frequency ("red") end boosted is called, very logically, "pink noise." If the other end were boosted we would undoubtedly call it "violet noise" (or lavender? orchid? mauve?).

DISTORTION

Our discussion of the various signals the hi-fi enthusiast deals with is incomplete without at least an acknowledgment of what can happen to the precious signal inside the black or hand-rubbed walnut boxes:

(a) **Bandwidth limitation**: If the passband of an amplifier cuts lows or highs, the signal is deteriorated. If the scratch filter reduces record surface noise, the overall effect may be improved but basically the signal itself is the poorer for it.

(b) **Nonuniform response**: Peaks and valleys within the passband also alter the signal waveshape.

(c) **Distortions in time**: If tape travels across the head at any other than its standard speed, the frequency components of the signal will be shifted up or down. If there are slow or fast fluctuations in that speed, wow or flutter is introduced and again we change the signal.

(d) **Dynamic distortion**: A compressor reduces the natural dynamic range of a signal and this is a form of distortion.

(e) **Nonlinear distortion**: If an amplifier, for example, is truly linear, there is a one-to-one relationship between input and output. Feedback helps to control nonlinear tendencies. The human ear is not linear. When a pure tone is impressed on the ear, harmonics can be heard. If two loud tones are presented simultaneously, sum and difference tones are generated **in the ear itself**; and these tones can be heard as can their harmonics. A cross-modulation test on an amplifier does essentially the same thing. If the amplifier (or the ear) were perfectly linear, no sum or difference tones or harmonics would be generated. The production within the black boxes of frequency components which were not present in the input signal is the result of nonlinear distortion.

(f) **Transient distortion**: Kick a bell and it rings. Apply a steep wavefront signal to an amplifier and it may ring a bit too. For this reason, signals such as piano notes are difficult to reproduce. Tone burst test signals are an attempt to explore the transient response characteristics of equipment, as are square waves.

When one considers all these terrible things that **could** happen to precious signals, one could become despondent, or jarred wide awake by the challenge all these potential pitfalls provide! We hope it will be the latter for the following chapters reveal other cruel acoustic fates waiting to pounce on unsuspecting signals.

4 RESONANCES IN ROOMS AND OTHER THINGS

Striking a bell causes it to vibrate at its "natural" frequency. A well-placed kick can set a suspended steel plate vibrating at a frequency determined by its physical makeup. In fact, a 3 by 6-foot steel plate has been used as an artificial reverberation device[10] but in this instance care was taken to operate at frequencies far removed from the natural frequency of the plate. The hardwood or metal bars of a xylophone vibrate at their natural frequencies determined by their size. Seismic disturbances in the earth are at frequencies far below the range of the human ear because of the great size of the earth. At California Institute of Technology the author heard a tape recording of such earth vibrations played back at a speed far higher than the recording speed. The earthquake had set the earth ringing like a bell!

Children are delighted to hear "the roar of the ocean" by holding a large conch shell to their ear. And who hasn't made a bottle "sound its A" by blowing across its mouth much as the jug player in a hillbilly band? Alas! Probably they do not realize the great acoustical significance of their actions! The spiral chamber of the great sea shell and the cavity within the jug are resonators springing to life in response to a bit of sound energy at the frequency at which the enclosed air is resonant. In the case of the conch shell those components of room noise near this natural frequency are selectively reinforced, producing the "ocean's roar." The stream of air across the mouth of the jug provides the energy which sets the air within singing the characteristic resonant tone of the jug.

RESONATORS

Hermann von Helmholtz (1821-1894) performed some interesting acoustical experiments with resonators functioning much as the conch shell and the jug. His resonators were a series of metal spheres of graded sizes, each fitted with a neck, appearing somewhat like the round bottom flask found in the chemistry laboratory. In addition to the neck there was another small opening to which he applied his ear. The resonators of different sizes resonated at different frequencies and by pointing the neck toward the sound under investigation he could estimate the energy at each frequency by the loudness of the sound in the different resonators.

There were numerous applications of this principle long before the time of Helmholtz. There is evidence that bronze jars were used by the Greeks in their open-air theaters to provide some artificial reverberation. A thousand years ago Helmholtz-type resonators were imbedded in church walls in Sweden with the mouths flush with the wall surface, apparently for sound absorption.[11] The walls of the modern sanctuary of Tapiola Church[12] in Helsinki, Finland, are dotted with slits in the concrete blocks (Fig. 4-1). These are resonator "necks" which open into cavities behind, together forming resonating structures. Energy absorbed from sound in the room causes each resonator to vibrate at its own characteristic frequency. Part of the energy is absorbed, part reradiated. The energy reradiated is sent in every direction, contributing to the diffusion of sound in the room. The resonator principle, old as it is, is continually appearing in modern, up-to-the-minute applications and we shall soon be applying it in the acoustic design of listening rooms and studios.

Bathroom Acoustics

Why is it that singing in the shower or tub is such a satisfying experience (to the singer at least)? Because here one's voice sounds richer, fuller and more powerful than anywhere else! The case of the bathroom baritone clearly illustrates the existence of resonant effects in a small room and the resulting reinforcement of sound at certain frequencies related to the dimensions of the room. Exciting the air in the bathroom at frequencies far removed from these characteristic frequencies results in weaker sounds, except at harmonics of these frequencies where the effect may be much like that at the fundamental frequencies.

Fig. 4-1. Helmholtz-type resonators built into the wall of the Tapiola Church in Helsinki, Finland. The slots and cavities behind act both as absorbers and diffusers of sound.

The man singing in the bathroom is, in a sense, inside an immense organ pipe but with one important difference; it is now a three- rather than an essentially one-dimensional system like the pipe. The hard walls of the bathroom are highly reflective. There is a fundamental frequency of resonance associated with the length, another with the width, and still another with the height of the bathroom. In the case of the cubical bathroom, all three fundamental frequencies coincide to give a mighty reinforcement to the baritone's voice at the fundamental and harmonic frequencies.

Resonance in a Pipe

The two ends of the pipe of Fig. 4-2 can be likened to two opposing walls of a listening room or studio. The pipe gives us a simple one-dimensional example to work with. That is, we can examine what happens between two opposite walls of a rectangular room without being bothered by the contributions of the other four surfaces. The pipe, closed at both ends and filled with air, is a resonator capable of sustained vibrations or oscillations. Air inside an organ pipe may be set to vibrating by blowing a stream of air across a lip at the edge of the pipe. It is simpler for us to place a small loudspeaker inside the pipe. Let us feed a sine wave signal to the loudspeaker and vary the frequency. If we drill a small hole in the pipe and place our ear against it we can hear the low-level tones radiated by the loudspeaker. As the frequency is increased, nothing much is noted until the frequency radiated from the loudspeaker coincides with the natural frequency of the pipe. At this frequency, f_0, modest energy from the loudspeaker is strongly reinforced and a relatively loud sound is heard at the earhole. As the frequency is increased, the loudness is again low until a frequency of $2f_0$ is reached, at which point another strong reinforcement is noted. Such resonant peaks can be detected at $3f_0$, $4f_0$...etc.

The graphs below the sketch of the pipe in Fig. 4-2 show how the sound pressure varies along the length of the pipe

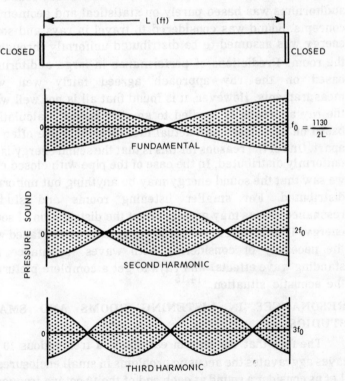

Fig. 4-2. A pipe closed at both ends helps us to understand how resonance occurs between two opposing wall of a listening room or studio. The distance between the walls determines the frequency of resonance, f_0.

under different conditions. A sound wave traveling to the right is reflected from the right plug and a sound wave traveling to the left from the left plug. The left-going waves react with the right-going waves creating, by superposition, a **standing wave** at the fundamental frequency of the pipe. Measuring probes inserted through tiny holes along the pipe could actually measure the high pressure near the closed ends and zero at the center, etc. Similar nodes (zero points) and antinodes (maxima) can be observed at $2f_0$, $3f_0$...etc., as shown in Fig. 4-2. The dimensions of a studio determine the characteristic frequencies of a room, much as though we had a north-south pipe, an east-west pipe and a vertical pipe, with the pipes corresponding to the length, width, and height of the room, respectively.

WAVES OR RAYS?

The classical work on reverberation of sound in auditoriums was based purely on statistical and geometrical concepts. Sound was considered to travel in rays and sound energy was assumed to be distributed uniformly throughout the room. Predictions of performance in large auditoriums based on the ray approach agreed fairly well with measurements. However, it is found that all is not well when the ray approach is applied to smaller rooms. Calculations based on ray theory and actual measurements are often far apart. One of the reasons for this is that the sound energy is not uniformly distributed. In the case of the pipe with closed ends we saw that the sound energy may be anything but uniformly distributed. For smaller listening rooms and studios, resonance effects may grossly affect the distribution of sound energy. Thus in the case of small enclosures we are faced with the necessity of considering both waves (resonance and standing wave effects) and rays to get a complete picture of the acoustic situation. [14,15]

RESONANCES IN LISTENING ROOMS AND SMALL STUDIOS

The fact that the human ear spans a tremendous 10 octaves aggravates the acoustic problems in small enclosures.[16] Let us consider a sound at each end of the 10-octave frequency range of the ear, one at 20 Hz and another at 20 kHz. At the low end a studio 28 feet long is resonant in this lengthwise mode at a frequency of 1130 / (2) (28) or 20 Hz. At this frequency and for a room of this size we have no alternative but to approach the acoustic problem from the standpoint of standing waves and resonances.

At 20 kHz, however, the wavelength is only about ⅝ inch which is very small compared to the 28-foot length of the room. While theoretically a standing wave having a wavelength of ⅝ inch could be maintained between the ends of the 28-foot room, practically speaking, small irregularities of the surfaces wound tend to diffuse the sound and thus destroy the resonant effect. Further, in trying to detect the standing wave there would be many problems because microphone diaphragms, eardrums, etc., are large compared to the wavelength and would thus tend to average rather than delineate resonant reinforcements and cancellations.

The transition from wave acoustics at the low end of the audible spectrum to ray acoustics at the high end is a very indefinite one. Very roughly we can say that for frequencies less than about 300 Hz, ray acoustics are of very limited utility. When the low frequency resonances in small rooms are adequately controlled in the general interests of sound quality, diffusion of sound is usually sufficient to give reasonable validity to reverberation calculations based on the ray approach.

5

STANDING WAVES IN LISTENING ROOMS AND SMALL STUDIOS

"Coloration" due to room resonances (normal modes) is often very serious with voice. Such colorations take the form of unnatural and monotonous emphasis of certain frequencies in the speaker's voice, giving it an unpleasant "roughness." Room colorations can also affect music but are more difficult to isolate because of the transient and nonrepetitive nature of most music. Also, the average listener knows better how a voice should sound than a given musical selection. Standing wave effects can accentuate particular musical notes as much as 8 or 10 dB.

In a listening room or studio, a **mode** is heard as a coloration of the desired sound if there is a tendency toward reinforcement at the modal frequency. If a certain modal frequency is isolated from its neighbors it is more likely to be audible. At the higher frequencies, say above about 300 Hz, the individual modes are seldom distinguishable, but below this frequency, isolated modes are common and often troublesome.

There is a tendency for acoustic faults of a room to be rejected by the binaural hearing mechanism of a person in that room (or hearing a stereo rendition of the sound which originated in that room). Because the degree of this binaural compensation is limited, it must not be used as an excuse for not correcting listening room and studio acoustics.

STANDING WAVE PATTERNS

The volume of air enclosed in a room is a complex vibratory system. To understand the acoustics of small listening rooms and studios is to understand this vibratory system and the large number of characteristic frequencies and standing waves associated with it.[17]

In Fig. 5-1 sound source S is located in the south wall of a rectangular room. There are three types of standing wave systems which can form. In sketch A an **axial** mode is represented in which reflections from only one pair of surfaces are involved, in this case the north and south walls. The east wall, west wall, floor, and ceiling are not involved because the waves travel parallel to them. The **tangential** waves are distributed parallel to two surfaces, parallel to floor and ceiling in sketch B, and involve reflections from the other four walls. The third vibration pattern involves reflections from all six surfaces. This mode is illustrated in sketch C and these are called **oblique** waves.

Fortunately for us, the axial modes are the more important for small listening rooms and studios having reasonably absorbent surfaces. Although we will concentrate our attention on the axial modes, we should never lose sight of the existence of tangential and oblique modes which, under some conditions, become troublesome.

The pipe resonance of Fig. 4-2 can be likened to one axial mode of a three-dimensional room. There are three axial modes in a rectangular enclosure such as that of Fig. 5-1. One of these, as illustrated in sketch A, involves the north and

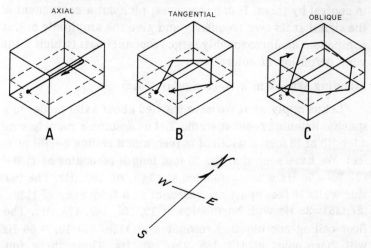

Fig. 5-1. There are three basic standing wave patterns which can form in a room. The axial mode (A) involves reflections from only one pair of surfaces. The tangential mode (B) involves two pairs of surfaces. The oblique mode (C) involves all six surfaces. The axial modes are of the greatest practical significance in small listening rooms and studios.

south walls, another the east and west pair of walls, and a third, the floor and ceiling. Even these three simplest modes become cumbersome as all three are acting simultaneously— and for this reason it is helpful to **study the standing waves of each axial mode separately.**

Mode Spacing

Colorations largely determine the quality of sound for a small studio or a listening room. The big task for us, then, is to determine which, if any, of the hundreds of modal frequencies in a room are likely to give us trouble.

It turns out that the spacing of the modal frequencies is a very important factor. Above, say, 300 Hz, the modal frequencies of a small room are so close together that they tend to merge harmlessly. At the lower audible frequencies, below about 300 Hz, their separation is greater, and it is in this region that problems can arise.

A good question would be "How close together must the modal frequencies be to avoid problems of coloration?" It has been found by mathematical analysis and experimental verification that if a modal frequency is separated more than about 20 Hz from its closest neighbor it will tend to be isolated acoustically.[18b] It will not be excited by its neighbors or held in control by them. It can, however, pick out a component of the signal at its own frequency and give the amplitude of this component an unreasonably large resonant boost (which is the very definition of coloration).

Standing Waves in a Rectangular Room

Let us apply what we have learned about axial modes to a specific listening room or studio. Let us assume a room having a length of 28 feet, a width of 16 feet, and a ceiling height of 10 feet. We have seen that the 28-foot length resonates at $1130 / (2)(28) = 20$ Hz with harmonics at 40, 60, 80, 100...Hz. The two side walls 16 feet apart are resonant at a frequency of $1130 / (2)(16) = 35$ Hz with harmonics at 70, 105, 140, 175...Hz. The floor-ceiling combination resonates at $1130 / (2)(10) = 56$ Hz with harmonics at 112, 168, 224, 280...Hz. These three fundamental frequencies and their corresponding series of harmonics are plotted in Fig. 5-2. There are 47 of these harmonics below 500 Hz.

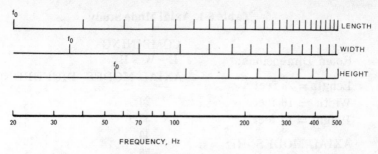

Fig. 5-2. The fundamental frequencies and harmonics of the three axial modes of a 16 x 28 x 10 foot room. The separation of these combined modal frequencies is an important criterion in predicting the quality of sound in a listening room or studio and in understanding problems in existing rooms.

Sounds complicated? This isn't even the half of it! The tangential and oblique modes and their harmonics add many, many more of these characteristic frequencies. Because of greater path lengths the tangential and oblique modes have lower fundamental frequencies and their amplitudes tend to be less because of the increased number of reflections.

Considering all three types of waves, there is an almost unbelievable number of axial, tangential, and oblique modes in a small studio. In the room just studied there are 1550 modes below 500 Hz and 12,000 below 1000 Hz![15, 19]

But do not be alarmed—the problems, though enormous in total, can be attacked piecemeal.

Because most signal colorations can be traced to axial modes, let us examine them in detail for the listening room or studio just considered. In Table 5-1 the fundamentals and harmonics of the three axial modes are tabulated on the left. On the right these modal frequencies are arranged in ascending frequency regardless of the mode or order of harmonic. We now inspect the differences between adjacent modal frequencies. If the spacings are greater than about 20 Hz we should be alert to possible problems. In Table 5-1, 20 Hz is the greatest spacing encountered, so from this standpoint everything would seem to be all right.

It is possible that zero spacings may also create problems. Zero spacing means that two or more modal frequencies are coincident, which might tend to overemphasize signal components at these frequencies. This would not be as serious as the cubical bathroom referred to earlier in which all three

61

Table 5-1. Axial Mode Study

Room Dimensions:

Length = 28 feet
Width = 16 feet
Height = 10 feet

AXIAL MODES - Hz

L			
fo	20	35	56
2fo	40	70	112
3fo	60	105	168
4fo	80	140	224
5fo	100	175	280
6fo	120	210	
7fo	140	245	
8fo	160	280	
9fo	180	315	
10fo	200		
11fo	220		
12fo	240		
13fo	260		
14fo	280		
15fo	300		

COMBINING L - W - H

AXIAL MODES	DIFFERENCE, Hz
20	15
35	5
40	16
56	4
60	10
70	10
80	20
100	5
105	7
112	8
120	20
(140)	0
(140)	20
160	8
168	7
175	5
180	20
200	10
210	10
220	4
224	16
240	5
245	15
260	20
(280)	0
(280)	0
(280)	20
300	15
315	

fundamentals as well as all harmonics are coincident. We note two coincident modes at 140 Hz (length $7f_0$ and width $4f_0$) and three at 280 Hz (length $14f_0$ and width $8f_0$ and height $5f_0$). We would thus be alerted to possible problems at these two frequencies. If they were detected by ear, tuned absorbers could be installed to control them.

Table 5-2 explores modal frequency spacings for a 16x14x10-foot studio or listening room. Note that the smaller room has fewer harmonics below 300 Hz. The two marginal spacings of 25 Hz mean only that coupling is weaker than desirable between the 80-105 Hz and the 175-200 Hz modes. As each mode is close to neighbors on the other side, no problems would be expected. We have a different problem at 280 Hz, however. Here three modes pile up and they are separated by 35 Hz from closest neighbors on either side. There would be a possibility that components of a voice signal near 280 Hz, for example, would be amplified acoustically, creating a serious coloration. Absorbers tuned sharply to this frequency would tend to tame these three coincident and isolated modal frequencies. In Fig. 5-3, however, we can see that there were few coloration problems at 280 Hz.

Standing Waves in Nonrectangular Rooms

If a listening room or a studio is built with one or two walls off square, will this eliminate the standing waves and all their problems? Certainly it would tend to diffuse the sound in the upper part of the audible spectrum where sound rays have meaning. Room resonance effects at the bass frequencies, however, would still be present as their identity is basically associated more with volume than with shape.

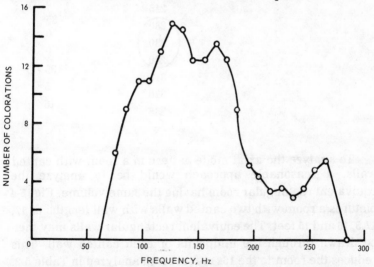

Fig. 5-3. A plot of 61 male voice colorations observed over a period of two years in BBC studios. Most fall in the 100 - 175 Hz band. Female voice colorations occur between 200 and 300 Hz. (After Gilford[18b]).

Table 5-2. Axial Mode Study

Room Dimensions:

Length = 16 feet
Width = 14 feet
Height = 10 feet

AXIAL MODES - Hz

	L	W	H
f_0	35	40	56
$2f_0$	70	80	112
$3f_0$	105	120	168
$4f_0$	140	160	224
$5f_0$	175	200	280
$6f_0$	210	240	336
$7f_0$	245	280	
$8f_0$	280	320	
$9f_0$	315		

COMBINING L — W — H

AXIAL MODES

	DIFFERENCE, Hz
35	
40	5
56	16
70	14
80	10
105	25 - ?
112	7
120	8
140	20
160	20
168	8
175	7
200	25 - ?
210	10
224	14
240	16
245	5
280	35 - ?
280	0
280	0
315	35 - ?
320	5
336	16

To analyze the axial mode pattern in a room with canted walls, a reasonable approach would be to analyze the equivalent rectangular room having the same volume. Fig. 5-4 pictures a room with two canted walls with wall lengths of 12, 17.5, 16 and 15 feet. The equivalent rectangular walls may then be drawn through the midpoints of each canted wall. This reduces the room to the 16x14-foot room analyzed in Table 5-2.

It is obvious that the nonrectangular wall of Fig. 5-4 would disturb two of the three axial modes, the floor-ceiling

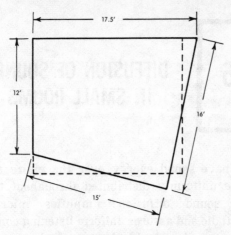

Fig. 5-4. Skewing the walls does not eliminate room resonances at the bass frequencies but it would make the sound field within more complex.

mode remaining essentially unaffected. However, the tangential and oblique modes would go on their merry ways, perhaps shifted somewhat but otherwise unaffected.

EXPERIMENTS WITH COLORATIONS

Any old ear can be offended by colorations caused by isolated modes but even a critical and trained ear needs some instrumental assistance in identifying and evaluating such colorations. The BBC Research Department made an interesting study. Observers listen to several men speaking in turn at a microphone in the studio under investigation, the voices being reproduced in another room over a high-quality system. Observers' judgments are aided by a selective amplifier which amplifies a narrow frequency band (10 Hz) to a level about 25 dB above the rest of the spectrum. The output is mixed in small proportions with the original signal to the loudspeaker, the proportion being adjusted until it is barely perceptible as a contribution to the whole output. Any colorations can then be clearly heard when the selective amplifier is tuned to the appropriate frequency.[18b]

In most studios tested this way, and we can assume they are well-designed, only one or two obvious colorations are found in each. Fig. 5-3 is a plot of 61 male voice colorations observed over a period of 2 years. Most fall between 100 and 175 Hz. Experience has shown that female voice colorations occur most frequently between 200 and 300 Hz.

6

DIFFUSION OF SOUND IN SMALL ROOMS

Our goal is to have sound energy over the entire audible frequency range uniformly distributed throughout the enclosure. Such sound diffusion simplifies microphone placement in a studio and assures uniform listening conditions in a listening room. A valuable by-product is that reverberation calculations based upon the assumption of completely diffuse sound give dependable answers. Of even greater significance, however, is the control of coloration of sound originating in a studio or being reproduced in a monitor or other listening room.

A highly simplified picture of a nondiffuse condition was seen in the pipe of Fig. 4-2. Considering only two parallel walls of the room analogy and only the fundamental frequency, f_0, the sound pressure is high near the walls and is zero at the center of the room. This is a very real physical effect, not just a theoretical concept. However, the $2f_0$ and the $3f_0$ and other harmonics of Fig. 4-2 which would be excited simultaneously by a complex voice or music signal would tend to obscure the pristine simplicity of the f_0 situation. The restoration of the other four walls and their attendant modes would further enormously complicate matters.

ROOM PROPORTIONS

If any two or three dimensions of a room are the same or if dimensions are in multiple relationship to each other, modal frequencies will coincide and there will be an unfortunate addition of their effects at these frequencies. For example, in a room whose length is 30 feet and height 10 feet we see the probability of this occurring. The axial modal frequencies of this room are tabulated in Table 6-1 out to $6f_0$. We see that 113, 226, and 339 Hz are common to all three dimensions. This room

Table 6-1. Coincident Room Resonances

AXIAL MODAL FREQUENCIES

	Height 10 ft	Width 20 ft	Length 30 ft
f_0	113Hz	56.5 Hz	37.6 Hz
$2f_0$	226	113	75
$3f_0$	339	169	113
$4f_0$	452	226	151
$5f_0$	565	282	188
$6f_0$	678	339	226

has many coincident frequencies in the north-south, east-west, and vertical modes simultaneously. Surely, this is not in the direction of achieving maximum sound diffusion.

There are preferred ratios of room dimensions (derived mathematically from wave acoustics) which will distribute the modal frequencies in the best way. In Fig. 6-1 is a plot showing an area of acceptable ratios growing out of the studies of Richard H. Bolt.[20,21] It is interesting to note in passing that most of the so-called "ideal" ratios urged by early acousticians fall outside this area. This may mean that many broadcasting studios constructed in the 1930s and 1940s utilizing these ratios are not optimized. On the other hand, it may also mean that ratios falling outside this area can yield reasonably acceptable results if other sound diffusing agents are used effectively. For checking proportions of studios or listening rooms already built, any ratios within the area should be considered acceptable except those having two or more dimensions in multiple relationship such as the example of Table 6-1. If the checkpoint falls outside the area, heroic efforts in obtaining diffusion by other means may be indicated.

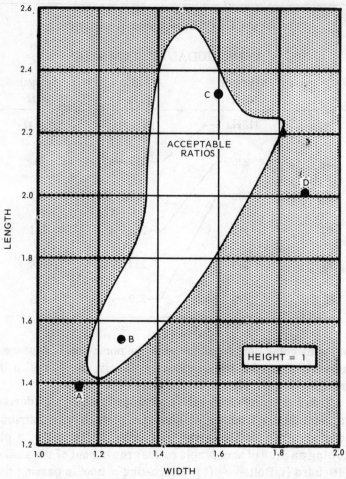

Fig. 6-1. Preferred ratios of room dimensions derived mathematically from wave acoustics to distribute modal frequencies in the best way. (After Bolt[20]).

A more recent computerized study by Ludwig W. Sepmeyer[22] seems to indicate that certain spots within and near Bolt's area are superior in distributing modes. Three of these preferred spots are indicated in Table 6-2. These are plotted on Fig. 6-1 as points A, B, and C. There is good reason to believe that the use of one of these three sets of dimension ratios will yield a better distribution of normal modes than just working within the acceptable area of Fig. 6-1. Table 6-3 expresses these ratios in terms of feet for ceiling heights of 8.5 and 10 feet.

Table 6-2. Preferred Room Dimension Ratios

	A	B	C
Height	1.00	1.00	1.00
Width	1.14	1.28	1.60
Length	1.39	1.54	2.33

Table 6-3. Preferred Room Dimensions

(Ceiling height 8.5 ft)			
	A	B	C
Height	8.5 ft	8.5 ft	8.5 ft
Width	9.7	10.9	13.6
Length	11.8	13.1	19.8
(Ceiling height 10 ft)			
	A	B	C
Height	10.0 ft	10.0 ft	10.0 ft
Width	11.4	12.8	16.0
Length	13.9	15.4	23.3

Let us run through a few examples of the use of Fig. 6-1. A 10x15x20 ft room would fall within the area (width 15/10=1.5; length 20/10 =2.0) but should be avoided because of the 2:1 ratio between length and ceiling height. For a ceiling height of 15 ft, the 1.6 width and 2.33 length ratios of preferred point C can be expressed in feet: width = (1.6)(15) = 24 ft; length =

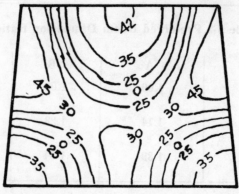

Fig. 6-2. Actual sound pressure contours measured in a trapezoidal room. (After Bolt et al.[13])

(2.33)(15)=35 ft. Let us check an existing 16x18 ft living room having an 8.5 ft ceiling. We divide 16 by 8.5 to obtain 1.88 width ratio and 18 by 8.5 to obtain a length ratio of 2.12. Plotting this point we locate point D, which falls outside the preferred area. Economic considerations might require continued use of this room, but the reason for diffusion problems is now obvious.

It should be mentioned that having favorable room proportions by no means assures good acoustics but it is a favorable start in the quest. Nor does an existing recording or listening room whose proportions lie outside the acceptable area mean unacceptable acoustics, but it does mean that more care is required in salvaging a less than optimum situation.

NONPARALLEL WALLS

If one of a pair of opposing walls is canted a few degrees from the rectangular position, successive reflections are moved to other surfaces rather than hitting the same spots each time. This tends to diffuse at least the high-frequency sound in the enclosure. At the lower frequencies, angling one or more walls does not eliminate room resonances. Fig. 6-2 shows measured sound pressures at a given frequency in a trapezoidal enclosure plotted as contour lines, the pressure being everywhere the same on a given contour.[23] Notice the three loops of zero pressure forming "valleys" separating high pressure "hills" typical of room resonances. At the higher frequencies the two canted walls would be more effective as diffusers. It would have been better to have two adjacent walls rather than two opposite walls canted, for this would have put one inclined wall in the north-south mode and one in the east-west mode.

CONCAVE SURFACES

Concave surfaces such as that in Fig. 6-3A tend to focus sound energy and consequently should be avoided because focusing is just the opposite of the diffusion we are seeking. The radius of curvature determines the focal distance; the flatter the concave surface the greater the distance at which sound is concentrated. Such surfaces often cause problems in microphone placement. Concave surfaces might produce some awe-inspiring effects in a "whispering gallery" where you can hear a pin drop 100 feet away, but they are to be avoided in listening rooms and small studios.

CONVEX SURFACES: THE POLY

One of the most effective diffusing elements, and one relatively easy to construct, is the polycylindrical diffuser (poly) which presents a convex section of a cylinder.[24,25] Three things can happen to sound falling on such a cylindrical surface made of plywood or hardboard; (1) the sound can be reflected and thereby dispersed as in Fig. 6-3B, (2) the sound can be absorbed, or (3) the sound can be reradiated. Such cylindrical elements lend themselves to serving as absorbers in the low-frequency range where absorption and diffusion are so badly needed in small rooms. The reradiated portion, because of the diaphragm action, is radiated almost equally throughout an angle of roughly 120 degrees as shown in Fig. 6-

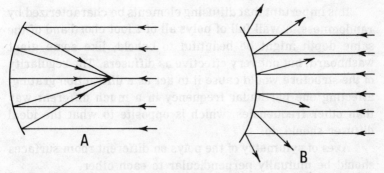

Fig. 6-3. Concave surfaces tend to focus sound, convex surfaces tend to diffuse it. Concave surfaces should be avoided if the goal is to achieve good diffusion of sound.

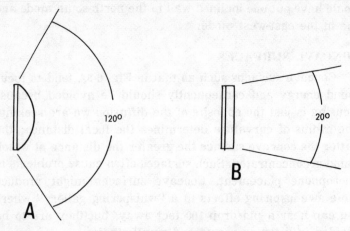

Fig. 6-4. (A) A polycylindrical diffuser reradiates energy not absorbed through an angle of about 120 degrees. (B) A similar flat element reradiates sound in a much smaller angle.

4A. A similar flat element reradiates sound in a much narrower angle, about 20 degrees. Therefore, favorable reflection, absorption, and reradiation characteristics favor the use of the cylindrical surface.

Some very practical polys and their absorption characteristics are presented in Chapter 8. The dimensions of such diffusers are not critical, although to be effective their size must be comparable to the wavelength of the sound being considered. The wavelength of sound at 1000 Hz is a bit over 1 foot, at 100 Hz about 11 feet. A poly element 3 or 4 feet across would be effective at 1000 Hz, much less so at 100 Hz. In general, poly base or chord lengths of 2 to 6 feet with depths of 6 to 18 inches meet most needs.

It is important that diffusing elements be characterized by randomness. A wall full of polys all of 2 foot chord and of the same depth might be beautiful to behold, like some giant washboard, but not very effective as diffusers. The regularity of the structure would cause it to act as a diffraction grating, affecting one particular frequency in a much different way than other frequencies, which is opposite to what the ideal diffuser should do.

Axes of symmetry of the polys on different room surfaces should be mutually perpendicular to each other.

The cylindrical surface is preferably made of ¼ or ⅜ in. plywood, although hardboard can also be used. The curved skin is wrapped around the ribs which serve not only as supports for the skin but also as bulkheads to break up the enclosed space. These smaller spaces discourage spurious modes of vibration from being set up and for this reason, random spacing of the ribs is recommended. The skin vibrates with considerable amplitude (easily felt with the fingertips) as sound fills the room. The relatively great absorption is the result of frictional losses in the fibers of the vibrating skin and, if used, the mineral fiber stuffing within. Thus the polycylindrical element is a good low-frequency absorber as well as a good diffuser.

Fig. 6-5. The construction of polys in a motion picture sound mixing studio which is basically a listening room. (Moody Institute of Science photo. Reprinted with permission from Journal of the Audio Engineering Society[26].)

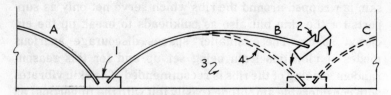

Fig. 6-6. The method of "stretching" the plywood or hardboard "skin" over the poly bulkheads shown in Fig. 6-5.

CONSTRUCTION OF POLYS

The **construction** of polycylindrical diffusers is reasonably simple.[26] Fig. 6-5 shows the framework for vertical polys mounted above a structure intended for a low-frequency slat absorber. Note particularly the variety of chord length and the random placement of bulkheads so that cavities will be of various volumes. Care should be taken to insure that each cavity is essentially airtight, isolated from nearby cavities by bulkheads and supporting framework fitting snugly to the wall. The bulkheads of each poly are carefully cut to the same radius on a bandsaw. Sponge rubber weatherstripping with an adhesive on one side is stuck to the edge of each bulkhead to insure a tight seal against the plywood or hardboard cover. If such precautions are not taken, annoying rattles and coupling between cavities can result.

At Moody Institute of Science, ⅛" tempered Masonite was used as the poly surface. A few hints can simplify the job of stretching the poly skin. In Fig. 6-6 slots of a width to fit the Masonite snugly are carefully cut the length of strips 1 and 2 with a radial saw. Let us assume that poly A is already mounted and held in place by strip 1 which is nailed or screwed to the wall. Working from left to right, the next job is to mount poly B. First the left edge of Masonite sheet B is inserted in the remaining slot of strip 1. The right edge of Masonite sheet B is then inserted in the left slot of strip 2. If all measurements and cuts have been accurately made, swinging strip 2 against the wall should make a tight seal over the bulkheads 3 and weatherstripping 4. Securing strip 2 to the wall completes poly B. Poly C is mounted in a similar fashion and so on to the end of the series of polys. The end result is shown in Fig. 6-7. Notice that the axes of symmetry of the polys on the side wall are perpendicular to those on the rear wall. If polys were used on the ceiling, their axes should be perpendicular to both the others.

It is quite practical and acceptable to construct each poly as an entirely independent structure rather than building them on the wall. Such independent polys can be spaced at will. In Chapter 8 the sound absorbing capability of polys will be discussed.

PLANE SURFACES

Geometrical sound diffusing elements made up of two flat surfaces giving a triangular cross section or of three or four flat surfaces giving a polygonal cross section may also be used. In general, their diffusing qualities are inferior to the cylindrical section.

DISTRIBUTION OF ABSORBING MATERIALS

It has long been known that the distribution of sound absorbing materials in patches throughout a room has a

Fig. 6-7. Finished poly array of Fig. 6-5 mounted on the wall above a low frequency absorber structure. Note that the axes of the polys on the rear wall are perpendicular to the axes of the polys on the other wall. (Moody Institute of Science photo. Reprinted with permission from Journal of the Audio Engineering Society.[26])

beneficial effect on the diffusion of sound. In U.S. practice, absorbing materials have been applied in geometrical designs to achieve both the desired diffusion and a pleasing esthetic effect.

The acoustics specialists of the British Broadcasting Corporation[27] compared two identical studios having approximately the same amounts and types of absorption. One of the studios also had diffusers in the form of cuboid protuberances fitted to the walls. Extensive reverberation and listening tests were inconclusive in establishing any acoustic superiority of one over the other. Adding the cuboid diffusers is expensive and of questionable esthetic value, hence the BBC now depends solely upon careful placement of different types of acoustic absorbers for their diffusion.

The secret of the BBC success is a modular approach to studio treatment. They use standard modules of 2 foot width, built to have several different acoustic characteristics (wideband absorbers, bass absorbers, etc.). The walls are completely covered with modules of the types required, and these modules are distributed over the various surfaces. The ceilings are covered with perforated metal trays concealing bass absorbers to offset the high-frequency absorption of the carpets. The BBC has achieved adequate diffusion in several hundred studios by careful distribution of the modules of different types.

CONTROL OF
INTERFERING NOISE

There are four basic approaches to reducing noise in a listening room or a recording studio:

(1) Locating the room in a quite place,

(2) Reducing the noise energy within the room,

(3) Reducing the noise output of the offending source,

(4) Interposing an insulating barrier between the noise and the room.

Locating a listening room or a studio away from outside interfering sounds is a luxury few can enjoy because of the many factors (other than acoustical) involved in site selection. If it is a hi-fi listening room and a part of a residence, due consideration must be given to serving the other needs of the family—at least if some degree of peace is to prevail. If the room in question is a recording or broadcast studio it is probably a part of a multipurpose complex and the noises originating from business machines, air conditioning equipment or foot traffic within the same building, or even sounds from other studios, may dominate the situation.

NOISE SOURCES, AND SOME SOLUTIONS

Protecting the room from street traffic noise is becoming more difficult all the time. It is useful to remember that doubling the distance from a noisy street or other sound source reduces the level of airborne noise approximately 6 dB. Shrubbery and trees may help in shielding from street sounds; a cypress hedge 2 ft thick gives about a 4 dB reduction.

The level of noise which has invaded a room by one means or another can be reduced by introducing sound-absorbing material into the studio. For example, if a sound level meter registers a noise level of 45 dB inside a studio, this level might be reduced to 40 dB by covering the walls with great quantities

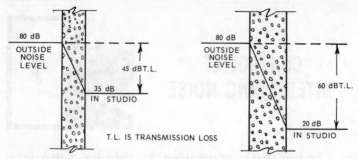

Fig. 7-1. The difference between the outside noise level and the desired noise level inside determines the required transmission loss of the wall.

of absorbing materials. Going far enough in this direction to reduce the noise significantly, however, would probably make the reverberation time too short. The control of reverberation must take priority. The amount of absorbent installed in the control of reverberation will reduce the noise level only slightly and beyond this we must look to other methods for further noise reduction.

Reducing the noise output of the offending source, if accessible and if possible, is the most logical approach. Traffic noise on a nearby street or airplanes overhead may be beyond control, but the noise output of a ventilating fan might be reduced 20 dB by the installation of a pliant mounting or the separation of a metal air duct with a simple canvas collar. Installing a carpet in a hall might solve a foot traffic noise problem or a felt pad a typewriter noise problem. In most cases working on the offending source and thus reducing its noise output is far more productive of results than corrective measures at or within the room in question.

As for terminology, we can say that a wall, for example, must offer a given "transmission loss" to sound transmitted through it as shown in Fig. 7-1. An outside noise level of 80 dB would be reduced to 35 dB by a wall having a transmission loss of 45 dB. A 60 dB wall would reduce the same noise level to 20 dB if no "flanking" or bypassing of the wall by other paths is present. We say that the wall "attenuates" the sound or that it "insulates" the interior from the outside noise. The walls, floor, and ceiling of the listening room or studio must give the required transmission loss to outside noises, reducing them to tolerable levels inside the room. Some of the principles involved in solving this problem will now be considered.

Noise may invade a studio or other room in the following ways:

(a) Airborne,

(b) Transmitted by diaphragm action of large surfaces, or

(c) Transmitted through solid structures, or

(d) Combination of all three.

AIRBORNE NOISE

A heavy metal plate with holes to the extent of 13 percent of the total area may transmit as much as 97 percent of the sound impinging on it. The amount of sound which can pass through a small crack or aperture in an otherwise solid wall is astounding. A crack under a door or a loosely fitting electrical service box can compromise the insulating properties of an otherwise excellent structure. Air-tightness is especially necessary to insulate against airborne noises.

NOISE CARRIED BY STRUCTURE

Unwanted sounds can invade an enclosure by mechanical transmission through solid structural members of wood, steel, or masonry. Air conditioner noises can be transmitted to a room by the air in the ducts, by the metal of the ducts themselves, or both. We are all familiar with the excellent sound-carrying capabilities of water pipes and plumbing fixtures.

It is very difficult to make a solid structure vibrate by airborne noise falling upon it because of the inefficient transfer of energy from tenuous air to a dense solid. On the other hand, a motor bolted to a floor, a slammed door or an office machine on a table with legs on the bare floor can cause the structure to vibrate very significantly. These vibrations can travel great distances through solid structure with little loss. In fact, with wood, concrete, or brick beams, longitudinal vibrations are attenuated only about 2 dB in 100 ft. Sound travels in steel about 20 times as far for the same loss! Although joints and cross-bracing members increase the transmission loss, it is still very low in common structural configurations.

NOISE TRANSMITTED BY DIAPHRAGM ACTION

Although very little airborne sound energy is transmitted directly to a rigid structure, airborne sound can set a wall to

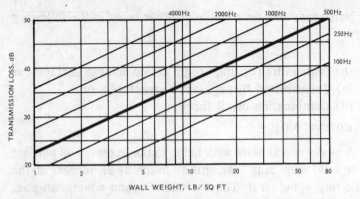

Fig. 7-2. The **mass** of the material in a barrier rather than the **kind** of material determines the transmission loss of sound going through the barrier. The transmission loss is also dependent on frequency although values at 500 Hz are commonly used in casual estimates. The wall weight is expressed in pounds per square foot of wall surface.

vibrating as a diaphragm and the wall, in turn, can transmit the sound through the interconnected solid structure. Such structure-borne sound may then cause another wall at some distance to vibrate, radiating noise into the room we are interested in protecting. Thus two walls interconnected by solid structure may serve as a coupling agent between exterior airborne noise and the interior of the listening room or studio itself.

SOUND-INSULATING WALLS

For insulating against outside airborne sounds, the general rule is the heavier the wall the better. The more massive the wall, the more difficult it is for sound waves in air to move it to and fro. Fig. 7-2 shows how the transmission loss of a rigid, solid wall is related to the density of the wall. The wall weight in Fig. 7-2 is expressed as so many pounds per square foot of surface, sometimes called the "surface density." For example, if a 10x10 ft concrete block wall weighs 2000 lb, the "wall weight" would be 2000 lb per 100 sq ft, or 20 lb per sq ft. The thickness of the wall is not directly considered.

From Fig. 7-2 we can see that the higher the frequency, the greater the transmission loss, or in other words, the better the wall is as a barrier to outside noises. The line for 500 Hz is made heavier than the lines for other frequencies as it is common to use this frequency for casual comparisons of walls of different materials. However, one should never forget that below 500 Hz the wall is less effective and for frequencies greater than 500 Hz it is more effective as a sound barrier.

The transmission losses indicated in Fig. 7-2 are based on the mass of the material rather than the kind of material. The transmission loss through a layer of lead of certain thickness can be matched by a plywood layer about 95 times thicker. But note that doubling the thickness of, say, a concrete wall, would increase the transmission loss only about 5 dB.

A discontinuous structure such as bricks set in lime mortar conducts sound less efficiently than a more homogenous material like concrete or steel. Therefore a brick wall would be more effective as a sound barrier than a con-crete wall of the same thickness. Unbridged air cavities between walls are very effective, but completely unbridged cavities are unattainable and only in the case of two separate structures, each on its own foundation, is this unbridged condition approached.

Porous Materials

Porous materials such as fiberglass (rockwool, mineral fiber) are excellent sound absorbers and good heat insulators but of limited value in insulating against sound. Using fiberglass to reduce sound transmission will help to a certain extent, but only moderately. The transmission loss for porous materials is directly proportional to the thickness traversed by the sound. This loss is about 1 dB (100 Hz) to 4 dB (3000 Hz) per inch of thickness for a dense, porous material (rockwool, density 5 lb/cu ft) and less for lighter material. Note that this direct dependence of transmission loss on thickness for porous materials is in contrast to the transmission loss for solid, rigid walls, which is approximately 5 dB for each doubling of the thickness.

Sound Transmission Classification(STC)

The solid line of Fig. 7-3 is simply a replotting of data from the mass law graphs of Fig. 7-2 for a wall weight of 10 lb per sq ft. If the mass law were perfectly followed, we would expect the transmission loss of a practical wall of this density to vary with frequency as shown by the solid line. Unfortunately, things are seldom this simple and we find that actual measurements of transmission loss on this wall might be more

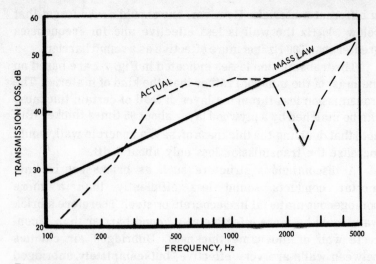

Fig. 7-3. Actual measurements of transmission loss in walls often deviate considerably from the mass law (Fig. 7-2) because of resonances and other effects.

like the broken line of Fig. 7-3. These deviations reflect resonance and other effects in the wall panel which are not included in the simple mass law concept.

Because of such commonly occurring irregularities, it would be of great practical value to agree on some arbitrary procedure of arriving at a single number which would give a reasonably accurate indication of the sound transmission loss characteristics of a wall. This has been done in a procedure specified by the American Society For Testing Materials in which the measured graph of a wall would be placed in a certain Sound Transmission Class (STC) by comparison to a reference graph (**STC contour**). The details of this procedure[30] are beyond the scope of this book but the results of such classification have been applied to walls of various types to be described for ready comparison. An STC rating of 50 dB for a wall would mean that it is better in insulating against sound than a wall of STC 40 dB. It is not proper to call STC ratings "averages" but the whole procedure is to escape the pitfalls of averaging dB transmission losses at various frequencies.

COMPARISON OF WALL STRUCTURES

Fig. 7-4 gives the measured performance of a 4" concrete block wall[28] as a sound barrier. It is interesting to note that plastering both sides increases the transmission loss of the wall from STC 40 dB to 48 dB. Fig. 7-5 shows a considerable improvement in doubling the thickness of the concrete block wall.[29] In this case the STC 45 dB is improved 11 dB by plastering both sides. In Fig. 7-6 we have the very common 2x4 frame construction with ⅝" gypsum board covering.[30] The STC 34 dB without fiberglass between is improved only 2 dB by filling the cavity with fiberglass material, a meager improvement which would probably not justify the added cost.

Fig. 7-7 describes a very useful and inexpensive type of wall of staggered stud construction. Here the inherently low coupling between the two independent wall diaphragms is further reduced by filling the space with fiberglass building material. Attaining the full STC 52 dB rating would require careful construction to insure that the two wall surfaces are truly independent and not "shorted out" by electrical conduits, outlet boxes, etc.

The last wall structure to be described is the double wall construction of Fig. 7-8. The two walls are entirely separate,

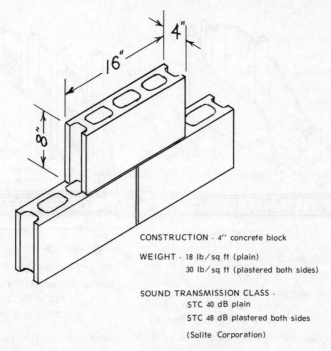

CONSTRUCTION - 4" concrete block

WEIGHT - 18 lb/sq ft (plain)
　　　　 30 lb/sq ft (plastered both sides)

SOUND TRANSMISSION CLASS -
　　　 STC 40 dB plain
　　　 STC 48 dB plastered both sides

(Solite Corporation)

Fig. 7-4. 4" concrete block.

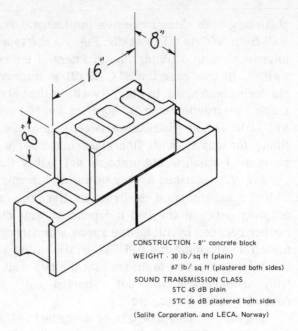

CONSTRUCTION - 8" concrete block
WEIGHT - 30 lb/sq ft (plain)
          67 lb/sq ft (plastered both sides)
SOUND TRANSMISSION CLASS
          STC 45 dB plain
          STC 56 dB plastered both sides
(Solite Corporation, and LECA, Norway)

Fig. 7-5. 8" concrete block.

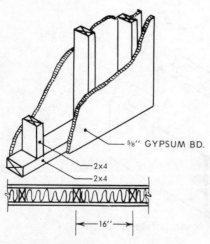

CONSTRUCTION - Standard stud partition
WEIGHT - 7.3 lb/sq ft
SOUND TRANSMISSION CLASS
          STC 34 dB without fibergl ss
          STC 36 dB with 3½" fiberglass
(Owens-Corning Fiberglas Corporation)[30]

Fig. 7-6. Standard stud partition.

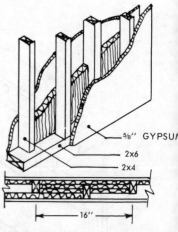

CONSTRUCTION - Staggered stud parti
WEIGHT - 7.2 lb/sq ft
SOUND TRANSMISSION CLASS
          STC 42 dB without fiberglas
          STC 46 to 52 dB with fibergl
(Owens-Corning Fiberglas Corporation

Fig. 7-7. Staggered stud partitior

84

each having its own 2x4 plate. Without fiberglass this wall is only 1 dB better than the staggered stud wall of Fig. 7-7 but by filling the inner space with building insulation, STC ratings up to 58 dB are possible.

Earlier in this chapter it was stated that porous sound absorbing materials are of limited value in insulating against sound. This is true when normal transmission loss is considered but in structures as those in Figs. 7-7 and 7-8, such porous materials have a new contribution to make in absorbing sound energy **in the cavity**. This can improve the transmission loss in some wall structures by as much as 15 dB, principally by reducing resonances in the space between the walls, while in others the effect is negligible. The low density mineral fiber batts commonly used in building construction are as effective as the high density boards and are much cheaper. Mineral fiber batts within a wall may also meet certain fire-blocking requirements in building codes.

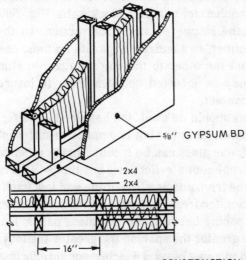

5/8" GYPSUM BD

2x4
2x4

16"

CONSTRUCTION - Double wall

WEIGHT - 7.1 lb/ sq ft

SOUND TRANSMISSION CLASS

STC 43 dB without fiberglass
STC 55 dB with 3½" fiberglass
STC 58 dB with 9" fiberglass

(Owens-Corning Fiberglas Corporation)[30]

Fig. 7-8. Double wall.

The staggered stud wall and the double wall, on the basis of mass alone, would yield a transmission loss of only about 35 dB (Fig. 7-2). The isolation of the inner and the outer walls from each other and the use of insulation within have increased the wall effectiveness by 10 or 15 dB.

DOUBLE WINDOWS

Between control room and studio a window is quite necessary and its sound transmission loss should be comparable to that of the wall itself. A well-built staggered stud or double wall may have an STC of 50 dB. To approach this performance with a window requires very careful design and installation.

A double window is most certainly indicated; a triple window adds little more. The mounting must minimize coupling from one wall to the other. One source of coupling is the window frame, another is the stiffness of the air between the glass panels. The plan of Fig. 7-9A is a practical solution to the double-window problem for concrete block walls. Fig. 7-9B is an adaptation to the staggered stud construction. In the latter there are, in effect, two entirely separate frames, one fixed to the inner and the other to the outer staggered stud walls. A felt strip may be inserted between them to insure against accidental contact.

Heavy plate glass should be used, the heavier the better. There is a slight advantage in having two panes of different thickness. If desired, one glass can be inclined to the other to control light or external sound reflections but this will have negligible effect on the transmission loss of the window itself. The glass should be isolated from the frame by rubber or other pliable strips. The spacing between the two glass panels has its effect, that is, the greater the spacing the greater the loss; but there is little gain in going beyond 8 inches nor serious loss in dropping down to 4 or 5 inches.

The absorbent material between the panes in the design of Fig. 7-9 discourages resonances in the air space. This adds significantly to the overall insulation efficiency of the double window and it should extend completely around the periphery of the window. If the double window of Fig. 7-9 is carefully constructed, sound insulation should approach that of an STC 50 dB wall but will probably not quite reach it. For the

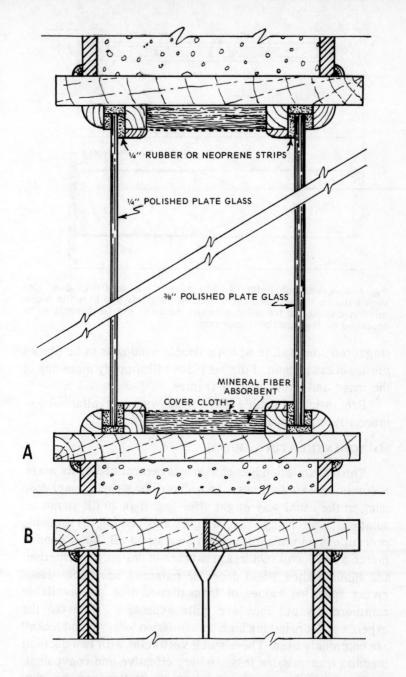

Fig. 7-9. Plans for double glass windows (A) for concrete block wall and (B) for staggered stud frame wall. Important features include the resilient mounting of the plate glass and the sound absorbing material around the periphery of the space between the two glass sheets.

Within the figure:
- ¼" RUBBER OR NEOPRENE STRIPS
- ¼" POLISHED PLATE GLASS
- ⅜" POLISHED PLATE GLASS
- MINERAL FIBER ABSORBENT
- COVER CLOTH
- A
- B

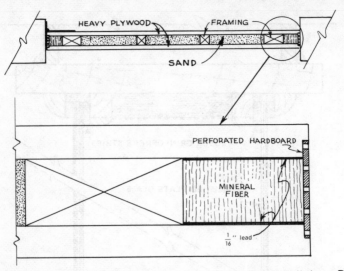

Fig. 7-10. A reasonably effective and inexpensive "acoustic" door. Dry sand between the plywood faces adds to the mass and thus the transmission loss. Sound traveling between the door and jamb tends to be absorbed by the absorbent door edge.

staggered stud wall in which a double window is to be placed the use of 2x8 instead of the 2x6 plate will simplify mounting of the inner and outer window frames.

Prefabricated double glass windows are available commercially, one of which is rated at STC 49 dB.

SOUND-INSULATING DOORS

The transmission loss of a door is determined by its mass, stiffness, and air-tightness. An ordinary household panel door hung in the usual way might offer less than 20 dB sound insulation. Increasing the weight and taking reasonable precautions on seals might gain another 10 dB, but a door to match a 50 dB wall requires great care in design, construction, and maintenance. Steel doors or patented acoustical doors giving specified values of transmission loss are available commercially but they are quite expensive. To avoid the expense of doors having high transmission loss, "sound locks" are commonly used. These small vestibules with two doors of medium transmission loss are very effective and convenient.

Doors having good insulating properties can be constructed if the requirements of mass, stiffness, and air-tightness are met. Fig. 7-10 suggests one inexpensive approach to the mass requirement, filling a hollow door with sand. Heavy plywood (¾ inch) is used for the door panels.

Achieving a good seal around a "soundproof" door can be very difficult. Great force is necessary to seal a heavy door. Wear and tear on pliant sealing strips can destroy their effectiveness, especially at the floor where foot-wear is a problem. The detail of Fig. 7-10 shows one approach to the sound leakage problem in which a very absorbent edge built around the periphery of the door serves as a trap for sound traversing the crack between door and jamb. This absorbent trap could also be imbedded in the door jamb. Such a soft trap could also be used in conjunction with one of the several types of seals.

Fig. 7-11 shows a do-it-yourself door seal which has proved reasonably satisfactory. The heart of this seal is a rubber or plastic tubing an inch or less in outside diameter with a wall thickness of about 3 / 32 inch. The wooden nailing strips hold the tubing to the door frame by means of a canvas wrapper. A raised sill is required at the floor if the tubing method is to be used all around the door (or another type of seal such as weatherstripping could be utilized at the bottom of the door). An advantage of tubing seal is that the degree of compression of the tubing upon which the sealing properties depend is available for inspection.

A complete door plan patterned after BBC practice[27] is shown in Fig. 7-12. It is based upon a 2" thick solid slab door and utilizes a magnetic seal such as used on refrigerator doors. The magnetic material is barium ferrite in a PVC (polyvinylchloride) rod. In pulling toward the mild steel strip, a good seal is achieved. The aluminum strip "C" decreases sound leakage around the periphery of the door.

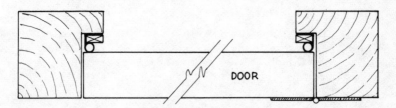

Fig. 7-11. A door may be sealed by compressible rubber or plastic tubing held in place by a fabric wrapper.

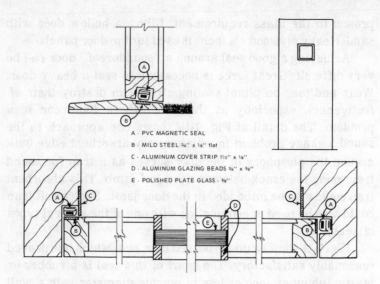

A - PVC MAGNETIC SEAL
B - MILD STEEL ¾" x ⅛" flat
C - ALUMINUM COVER STRIP 1½" x ⅛"
D - ALUMINUM GLAZING BEADS ¾" x ⅜"
E - POLISHED PLATE GLASS - ⅝"

Fig. 7-12. A BBC door design utilizing magnetic seals of the type used on refrigerator doors. (After Brown[27])

It is possible to obtain a very slight acoustical improvement and, to some at least, an improvement in appearance by padding both sides of a door. A plastic fabric over 1" foam rubber sheet may be "quilted" with upholstery tacks. (Refer to Figs. 14-1 through 14-5 for an example.)

NOISE AND ROOM RESONANCES

Room resonances can affect the problem of outside noise in a studio. Any prominent modes persisting in spite of acoustic treatment make a room very susceptible to interfering noises having appreciable energy at these frequencies. In such a case a feeble interfering sound could be augmented by the resonant effect to a very disturbing level.

SOUND ABSORBERS

8

The absorption of sound is essentially the dissipation of sound energy as heat. As a million people talking simultaneously for over 2 hours would produce only enough heat to brew a cup of tea, the absorbed sound in a small listening room or studio will neither solve a heating problem in cold climates nor cause a refrigeration problem in the tropics!

The sound absorbed at boundary surfaces can be accounted for primarily by (a) porosity and (b) flexural vibrations. In addition there is a family of structures involving resonance phenomena whose sound-absorbing qualities may also be influenced by porosity and panel flexure. Practically all of the welter of sound-absorbing materials and structures available today fall into one of these three categories.[31,32,33]

POROUS ABSORBERS

Sound is absorbed in the tiny spaces of porous materials such as felt, rugs, carpets, drapes, and a number of fibrous batts and boards made of cellulose or mineral fibers. Most of the proprietary acoustic materials are of the porous, fibrous type. The sound absorbed by bricks and concrete blocks is due chiefly to porosity.

The absorption efficiency of materials depending upon the trapping of sound in tiny pores can be seriously impaired by painting, as paint fills the absorbing pores. Acoustic tile is available with a factory applied painted surface which minimizes the problem. Under certain conditions a painted surface can act as a diaphragm which actually tends to make up for the loss of absorption through porosity.

In the first radio broadcasting studios the acoustic treatment was a blind and overenthusiastic use of drapes and carpets. Soon tiles of cellulose fiber, usually with surface

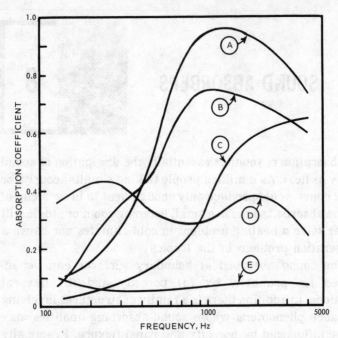

Fig. 8-1. Sound absorption coefficients of typical porous materials which absorb best in the high frequencies.[31]
(A) Simpson Pyrotect standard drilled tile ½" thick.
(B) Medium weight (14 oz per sq yd) velour draped to half area.
(C) Heavy carpet laid on concrete without padding.
(D) Coarse concrete blocks, unpainted, and (E) the same blocks painted.

perforations, augmented the drapes and carpets. The sad thing about the situation was that all of these materials absorbed well in the treble frequencies but poorly in the bass. As a result, room modes were allowed to do their dirty work of distorting and accentuating the bass with little restraint while the music overtones were soaked up and the intelligibility of voice was reduced. It is difficult to imagine a more unfortunate combination, and all with the finest intentions!

Today we know better. Or do we? When you see the walls and ceiling of a studio or listening room covered with cellulose tile, we're right back to 1930! Acoustic tile, rugs, and drapes are excellent acoustic materials, if used correctly. We shall study their characteristics so that we will be able to know what to use, where and how much.

Several sound-absorbing materials depending upon porosity for their effect have absorption coefficients which vary with frequency as shown in Fig. 8-1. Graph A is for typical perforated cellulose tile ½" thick and cemented to a plaster or concrete wall surface. The absorption is very high in the 600-4000 Hz range but poor at lower frequencies. Graph B is for medium weight velour (14 oz per sq yd) draped to half area. Graph C is for heavy carpet laid on concrete without padding. The drapes and carpets share the general characteristic of good absorption in the highs and poor in the lows.

Concrete blocks would seem to have little in common with cellulose tile, drapes, or carpets, but concrete blocks also absorb sound by porosity. Graph D gives the sound-absorption coefficients for coarse concrete blocks. Graphic proof that the sound is really absorbed by the surface pores of the concrete blocks is given in graph E which is for the same blocks but with the surface painted. Painting the surface of the blocks reduces the absorption from the 30-40 percent region to 5-10 percent.[31] So do not paint them.

There is a variable aspect to drapes which affects their absorption, that is, the fullness with which they are hung.[34] Fig. 8-2 shows how the absorption of one cotton drapery material varies with the fullness. "Draped to 7/8 area" means that it is almost flat, having only that fullness which moving from 8/8 area (full area of the material) to 7/8 of the

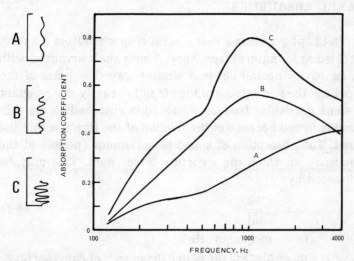

Fig. 8-2. The more drapery is folded, the higher is the absorption coefficient. Material: cotton cloth, 14.7 oz / sq yd.
(A) Draped to 7/8 area, or almost flat.
(B) Draped to 3/4 area.
(C) Draped to 1/2 area. (After Mankovsky17)

area would provide. "Draped to 1/2 area" means that the drape with folds occupies only half the area of the material itself. The folds of the drape are seen to be particularly effective in absorbing sound of the higher frequencies.

The location of the drapes in the room and the distance between the drape and the wall influence the absorption of sound.[35] As sound is reflected from the hard wall, the particle velocity is maximum at a distance of a quarter wavelength and odd multiples of a quarter wavelength from the wall. If cloth is hung at this distance from the wall, the high particle activity results in high absorption. For example, the wavelength of sound at 1000 Hz is about 1.1 ft (see Fig. 1-7). Thus for 1000 Hz, drapery material at a distance of about 3" from the wall and 3, 5, 7...etc. times this distance would thus yield maximum absorption. Because of this effect one would expect ups and downs in absorption for different frequencies but these fluctuations tend to be obscured by the method of determining the absorption coefficients. Suffice it to say that draperies are reputable acoustic elements which have the advantage of being adjustable by being extended and retracted at will.

PANEL ABSORBERS

In Chapter 1 we saw that a mass suspended from a spring vibrated at its natural frequency. Thin panels arranged with an air cavity behind act in a similar way. The mass of the panel and the springiness of the air in the cavity are resonant at some particular frequency. Sound is absorbed as the thin panel is flexed because of the friction of the fibers within the panel. The absorption of sound is maximum (peaks) at the frequency at which the structure is resonant. This may be estimated by:

$$f_0 = \frac{170}{\sqrt{md}} \tag{8-1}$$

where
f_0 = frequency, Hz
m = mass of the panel, lb per sq ft of panel surface
d = depth of air space, inches

To make it easy to use, equation 8-1 has been plotted in Fig. 8-3 for some commonly used materials. From the graph we see, for example, that ⅛" plywood furred out from a solid wall by 2x4s on edge resonates at about 145 Hz. Masonite or other hardboard of ¼ inch thickness and spaced out from the wall with 2x6s on edge resonates near 70 Hz.

Normally, flat panels used as acoustic elements would be spaced out from the wall with mineral wool in the cavity. This more than doubles the peak absorption coefficient. In Fig. 8-4 the absorption characteristics of three panel configurations are shown. In graph A is the simple case of 3/16" plywood panels on 2" battens.[32] If we refer to Fig. 8-3 and exercise a bit of judgment in reading midway between the 1/8 and 1/4" plywood graphs, we estimate that this structure is resonant at about 175 Hz. The peak coefficient is about 0.4 which is about

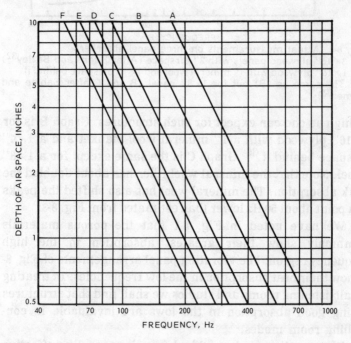

Fig. 8-3. A graphical presentation of Equation 1 relating the mass of the panel, the depth of the airspace, and the frequency of resonance of the panel absorber.
(A) Mass=0.1 lb per sq ft
(B) Mass=0.2 lb per sq ft
(C) Mass=0.37 lb per sq ft (⅛" plywood)
(D) Mass=0.55 lb per sq ft (⅛" hardboard)
(E) Mass=0.74 lb per sq ft (¼" plywood)
(F) Mass=1.1 lb per sq ft (¼" hardboard)
Example: ¼" plywood with airspace depth of 3.75 inches resonates at about 100 Hz.

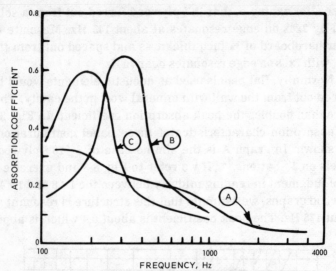

Fig. 8-4. Actual measurements on three panel absorbers.
(A) 3/16" plywood panels with 2" airspace (After Evans and Bazley[32])
(B) 1/16" plywood with 1" mineral wool and 1/4" airspace behind it
(C) The same as (B) but for 1/8" panel (B and C after Sabine and Ramer[36])

as high as one can expect for such structures. Graph B is for 1/16" plywood with a 1" mineral wool blanket and a 1/4" airspace behind it.[36] Graph C is the same except for a 1/8" panel. Note that the mineral wool filler has about doubled the peak absorption. The mineral wool has also shifted the peaks to a point about 50 Hz lower than estimated from Fig. 8-3.

We have noted in Fig. 8-1 that the porous materials commonly show their greatest absorption in the high frequency region. The vibrating panel arrangements of Fig. 8-4 show their best absorption in the low frequencies. In treating small listening rooms and studios we shall find that structures giving good absorption in the lows are invaluable in controlling room modes.

Flat paneling can be highly decorative as well as effective in absorbing sound in the bass region. In fact, the excellent acoustics of some famous music rooms in Europe can be traced to the rich hardwood panels covering the walls.

POLYS: WRAPAROUND PANELS

Flat paneling in a room may brighten an interior decorator's eye and do some good acoustically, but wrapping a thin plywood or hardboard skin around some semicylindrical bulkheads can provide some very attractive features, as we have seen in Chapter 6. These polycylindrical elements (polys) are rather out of fashion now after a strong run of popularity in radio broadcasting, listening and recording studios, and motion picture scoring stages in the 1940s and 1950s. Visually, they are rather overpowering, which can be good or bad depending upon the effect one wants to achieve. With polys it is acoustically possible to achieve a good diffuse field along with liveness and brilliance, factors tending to oppose each other in rooms with flat surfaces.

One of the problems of using polys has been the scarcity of published absorption coefficients. The Russian acoustician, V.S. Mankovsky, has taken care of that in his recent book.[17] As expected, the larger the chord the better the bass absorption. Above 500 Hz there is little significant difference between the different sizes of polys.

The overall length of polys is rather immaterial, ranging in actual installations from the length of a sheet of plywood to the entire length, width, or height of a studio. It is advisable, however, to break up the cavity behind the poly skin with randomly spaced bulkheads. The polys of Fig. 8-5 incorporate such bulkheads which break up the 5- to 6-foot polys into sections having approximate lengths as follows:

POLY SIZE			
A	B	C	D
20	18	16	14 inches (Top)
16	14	12	10
12	10	8	8
8	8	10	6
12	10	12	8
			10
			12 (Bottom)

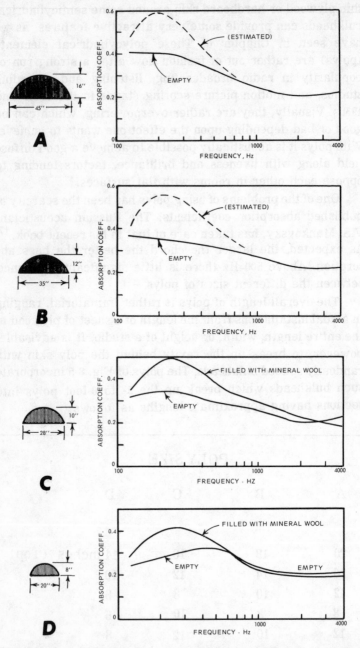

Fig. 8-5. Measured absorption of polycylindrical diffusers of various chord and height dimensions. In C and D, graphs are shown for both empty polys and for polys filled with mineral wool. In A and B only empty poly data are available; the broken lines are estimated absorption when filled with mineral wool. (After Mankovsky[17])

Should the polys be empty or filled with something? Mankovsky again comes to our rescue and shows us the effect of filling the cavities with mineral (glass) wool. Figs. 8-5C and 8-5D show the increase in bass absorption resulting from filling the cavities with mineral wool. If needed, this increased bass absorption may be easily achieved simply by filling the polys with glass wool. If the bass absorption is not needed the polys may be used empty. The great value of this adjustable feature will become more apparent as we get into the actual acoustical design of listening rooms and studios in Chapter 9.

PERFORATED PANEL ABSORBERS

Perforated hardboard or plywood panels spaced from the wall constitute a resonant type of sound absorber.[37,38] Each hole acts as the neck of a Helmholtz resonator and the share of the cavity behind "belonging" to that hole is comparable to the cavity of the Helmholtz resonator. In fact, we can view this structure as a host of **coupled resonators**. If sound arrives perpendicular to the face of the perforated panel, all the tiny resonators are in phase. For sound waves striking the perforated board at an angle, the absorption efficiency is somewhat decreased. This loss can be minimized by sectionalizing the cavity behind the perforated board with an egg crate type of divider of wood or corrugated paper.

The frequency of resonance of perforated panel absorbers backed by a subdivided air space is given approximately by:

$$f_0 = 200\sqrt{\frac{p}{dt}} \qquad (8\text{-}2)$$

where f_0 = frequency, Hz

p = percentage of open area
= hole area divided by panel area x 100

t = effective hole length, inches, with correction factor applied
= (panel thickness) + (0.8)(hole diameter)

d = depth of air space, inches

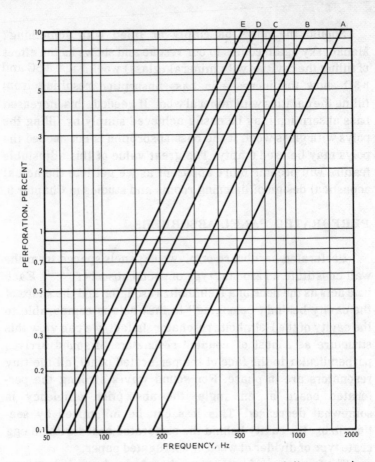

Fig. 8-6. A graphical presentation of Equation 8-2 relating percentage perforation of perforated panels, the depth of airspace, and the frequency of resonance. These graphs are for panel thickness of 3/16".
(A) 1" furring lumber (The lines are drawn to correspond to furring lumber which is finished, e.g., the line for 8" is actually 7-3/4" airspace)
(B) 2", (C) 4" and (D) 8".

Eq. 8-2 is true only for circular holes. This information is presented in graphical form in Fig. 8-6 for a panel thickness of 3/16". Common hardboard perforated with holes 3/16" in diameter spaced 1" on centers has 2.75 percent of the area in holes. If this perforated panel is spaced out from the wall by 2x4s on edge, the system resonates at about 420 Hz and the peak absorption would appear near this frequency.

In commonly available materials the holes are so numerous that resonances at only the higher frequencies can be obtained with practical air space depths. For example,

Johns-Manville Transite (a 3/16" asbestos board) has a 10.5 percent perforation. On 2x4s on edge it resonates at about 800 Hz. To obtain much-needed low frequency absorption you may have to drill your own! Drilling 7/32" holes 6" on centers comes out around 1.0 percent perforation. With zero perforation percentage you are right back to the solid panels of Fig. 8-3!

Fig. 8-7 shows the effect of varying the hole area from 0.18 percent to 8.7 percent in a structure of otherwise fixed dimensions. The plywood is 5/32" thick perforated with 3/16" holes except for the 8.7 percent case in which the hole diameter is about 3/4". The perforated plywood sheet is spaced 4" from the wall and the cavity is half filled with mineral wool and half is air space.[17]

Fig. 8-8 is identical to Fig. 8-7 except that the perforated plywood is spaced 8" and the mineral wool of 4" thickness is mounted in the cavity. The general effect of these changes is a substantial broadening of the graphs and a minor lowering of the peak frequency.

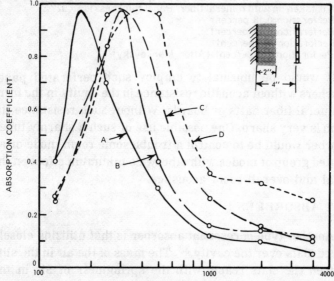

Fig. 8-7. Actual absorption measurements on perforated panel absorbers of 4" airspace, half filled with mineral wool and for panel thickness of 5/32".

(A) Perforation 0.18 percent
(B) Perforation 0.79 percent
(C) Perforation 1.4 percent and (D) 8.7 percent. Note that the presence of the mineral wool shifts the frequency of resonance considerably from the theoretical values of Equation 8-2 and Fig. 8-6. (After Mankovsky[17])

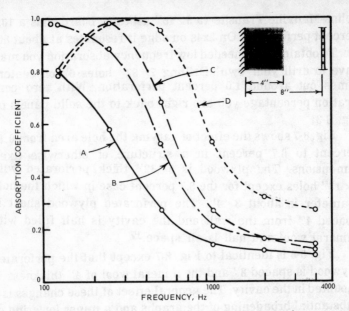

Fig. 8-8. Actual absorption measurements on the same perforated panel absorbers of Fig. 8-7 except that the airspace is increased to 8″, half of which is taken up with mineral fiber. Panel thickness is 5/32″.
(A) Perforation 0.18 percent
(B) Perforation 0.79 percent
(C) Perforation 1.4 percent
(D) Perforation 8.7 percent (After Mankovsky[17])

It would be unusual to employ such perforated panel absorbers without acoustic resistance in the cavity in the form of mineral fiber batts or boards. Without such resistance the graph is very sharp. One possible use of such a sharply tuned absorber would be to control a troublesome room mode or an isolated group of modes with otherwise minimum effect on the signal and overall room acoustics.

SLAT ABSORBERS

Another type of resonant absorber is that utilizing closely spaced slats over the cavity.[38] The mass of the air in the slits between the slats reacts with the springiness of air in the cavity to form a resonant system, again comparable to the Helmholtz resonator. The mineral fiber board usually introduced behind the slits acts as a resistance, broadening the peak of absorption. The narrower the slits and the deeper the cavity, the lower the frequency of maximum absorption. Unfortunately, the computation of the resonant frequency of slit-cavity resonators is a bit too complex for this presentation. A cross section of a typical slit resonator which peaks at about 150 Hz is shown in Fig. 8-9.

Suggestion: if slats are mounted vertically, it is recommended that they be finished in a dark color conforming to the shadows of the slits to avoid some very disturbing "picket-fence" optical effects!

MIDRANGE ABSORBERS

Fig. 8-10 illustrates midrange absorption available by mounting mineral fiber boards on 1'' strips (1x3s mounted flat on 12'' centers) to give a 1'' air space between the wall and the fiberboard.[39] Graph A is for 1½'' board of 9 lb per cu ft density while graph B is for ½'' fiberboard of 2 lb per cu ft density.

The person designing a listening room or a small studio for certain reverberation characteristics will find a wealth of

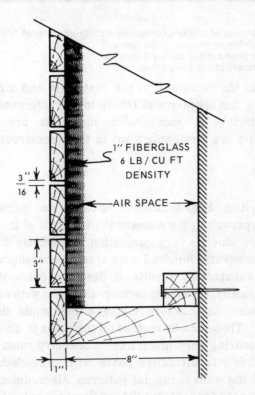

Fig. 8-9. A typical slat absorber giving peak absorption at about 150 Hz.

103

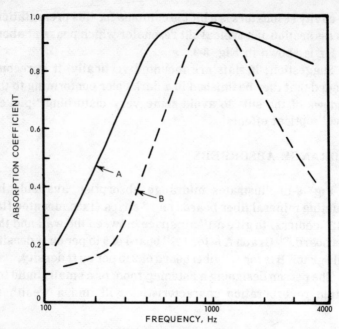

Fig. 8-10. Midrange absorbers composed simply of mineral fiber boards mounted on strips to give a 1" airspace.[39]
(A) 1-½" fiber board of 9 lb per cu ft density
(B) ½" fiber board of 2 lb per cu ft density

resources at his command in the materials and structures described in this chapter and others in the reference books. These materials or comparable materials are readily available and are straightforward in their construction.

MODULES

The British Broadcasting Company has pioneered a modular approach to the acoustical treatment of their small voice studios which is very interesting.[27] Because they have applied it to several hundred such studios economically and with very satisfactory results, it deserves our critical attention. Basically, the idea is to cover the walls with standard-sized modules, say, 2x3 ft having a maximum depth of, perhaps, 8". These can be framed on the walls to give a flush surface appearing very much like an ordinary room or they may be made into attractive boxes with grill cloth covers mounted on the walls in regular patterns. All modules may be made to appear identical, but the similarity is only skin deep.

There are commonly three or perhaps four different types, each having its distinctive contribution to make acoustically. Fig. 8-11 shows the radically different absorption characteristics obtained by merely changing the covers of the standard module. This is for a 2x3 ft module having a 7" air space and a 1" semirigid glass fiber board of 9 to 10 lb per cu ft density inside. The wideband absorber has a highly perforated cover (25 percent or more of the area is holes) or no cover at all, yielding essentially complete absorption down to about 200 Hz. Even better low-frequency absorption is possible by breaking up the air space with egg crate type dividers of corrugated paper to discourage unwanted resonance modes. A cover ¼" thick with 5 percent of the area in holes peaks in the 300-400 Hz range. A true bass absorber is obtained with a low perforation cover (0.5 percent of the area in holes). If essentially neutral modules are desired, they can be covered with ⅜ or ¼" plywood which would give relatively low absorption with a flat peak about 70 Hz. Using these three or four modules as acoustic building blocks, the desired effect can be

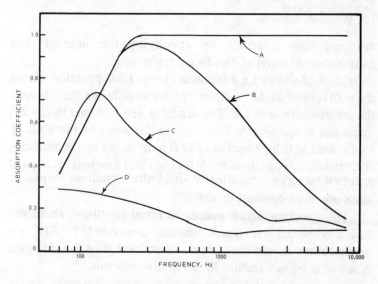

Fig. 8-11. Module absorbers having a 7" airspace and a 1" semirigid glass fiber board of 9 to 10 lb per cu ft density behind the perforated cover.
(A) No perforated cover at all (or at least over 25 percent perforation)
(B) 5 percent perforated cover
(C) 0.5 percent perforated cover
(D) ⅜" plywood cover essentially to neutralize the module
(A, B and C adapted from ref. 33)

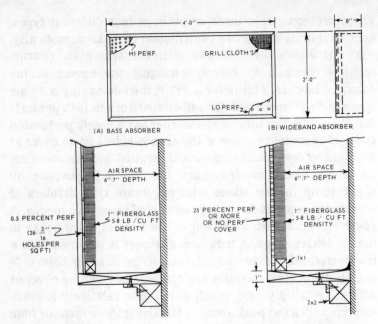

Fig. 8-12. Plan for a practical module absorber utilizing the wall as the bottom of the module.
(A) Bass absorber
(B) Wideband absorber

designed into a studio by specifying the number and distribution of each of the basic types used.

Fig. 8-12 shows an adaptation from BBC practice where the wall is used as the "bottom" of the module box. In this case the module size is 2x4 ft. The modules are fastened to the 2x2 mounting strips which, in turn, are fastened to the wall. A studio wall 10 ft high and 23 or 24 ft long might use 20 modules of distributed types, four modules high and five long. It is good practice to have acoustically dissimilar modules opposing each other on opposite walls.

The question which comes to mind is, "How about diffusion of sound with such modular treatment?" BBC experience has shown that careful distribution of the different types of modules results in adequate diffusion.[27]

PLACEMENT OF MATERIALS

The application of sound absorbing materials in random patches has already been mentioned as an important means of achieving diffusion. Other factors than diffusion might influence placement. If several types of absorbers are used, it is desirable to place some of each type on ends, sides, and ceiling so that all three axial modes (longitudinal, transverse, and vertical) will come under their influence. In rectangular rooms it has been demonstrated that absorbing material placed near corners and along edges of room surfaces is most effective. In speech studios, some absorbent effective at high frequencies should be applied at head height on the walls. In fact, material applied to the lower portions of high walls may be as much as twice as effective as the same material placed elsewhere. It is rather an obvious conclusion that untreated surfaces should never face each other.

Winston Churchill once remarked that as long as he had to wear spectacles he intended to get maximum cosmetic benefit from them. So it is with placement of acoustic materials. After the demands of acoustic function have been met, every effort should be made to arrange the resulting patterns, textures, and protuberances into an esthetically pleasing arrangement, but do not let your priorities be shifted!

9

REVERBERATION AND HOW TO COMPUTE IT

If you push the accelerator pedal of your automobile down to a certain point and hold it there, the car accelerates to a certain speed, and if the road is smooth and level this speed will remain constant. With this accelerator setting the engine produces just enough torque to overcome all the frictional losses and a balanced or steady-state condition results. So it is with sound in a room. A loudspeaker is arranged to emit white noise into a room. As the switch is closed the sound quickly builds up to a certain level as shown in Fig. 9-1. This is the steady-state or equilibrium point at which the sound energy radiated from the loudspeaker is just enough to supply all the losses at the boundaries of the room. If the switch is opened, it takes a finite length of time for the sound level to decay to inaudibility. This "hanging-on" of the sound in a room after the exciting signal has been removed is called **reverberation** and it has a very important bearing on the acoustic quality of the room.

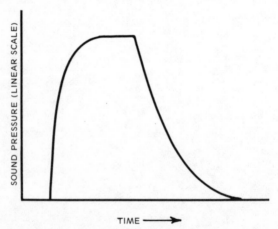

Fig. 9-1. Rise and decay of sound in a room. The linear sound pressure scale yields the familiar exponential decay.

Recordings made outdoors or in an echo-free room have a dead quality. Sound may spread in all directions from the source but only that traveling toward the microphone will be recorded under these circumstances. If the same source, microphone, and recording system are taken indoors, reflections from the walls, ceiling, and floor give the recorded sound quite a different quality. Controlled reverberation, whether natural or artifical, can enhance the quality of voice and music being recorded or reproduced but too much or too little reverberation can create problems. In the design of small listening rooms or studios, reverberation must be considered along with diffusion of sound and the intrusion of unwanted sounds from the outside.

REVERBERATION AND NORMAL MODES

A room is a complex vibratory system, as we have seen, with a whole array of characteristic frequencies as shown in Tables 5-1 and 5-2. If the loudspeaker emits a pure 280 Hz tone, the 280 Hz modes of Table 5-2 will be strongly excited. As the loudspeaker switch is opened, the energy of this mode will die away. We can probably force the 280 Hz mode to vibrate at 275 Hz by shifting the tone on the loudspeaker to this frequency. As soon as the loudspeaker signal is cut, however, the stored energy will probably decay at its own natural frequency of 280 Hz. This mode vibrates at 275 Hz as a "forced" vibration but decays at its own natural 280 Hz as a "free"vibration. Isolated modes are especially subject to such frequency shifting, a form of distortion. The effect is minimized if all modes are closely spaced. The isolated pile-up of modes at 280 Hz in Table 5-2 would be very susceptible to such problems.

Voice and music signals are characterized by energy constantly shifting in frequency in a highly transient time pattern. It is impossible for us to grasp anything but a very general picture of the extremely complicated excitation of the many room modes by these complex and constantly shifting signal components.

Using statistical methods, the experts have lumped all the axial, tangential, and oblique modes together and have come up with a relatively simple picture of the overall reverberation characteristics of a listening room or studio. Fortunately for

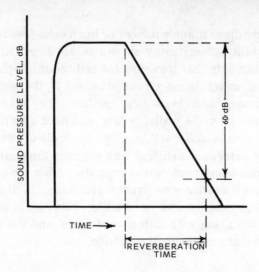

Fig. 9-2. Rise and decay of sound in a room with a logarithmic vertical scale, sound pressure level in dB. The exponential decay is now a straight line, convenient for measuring reverberation time which is defined as the time for sound to decay 60 dB.

us, this simplified picture is confirmed by listening tests. After all, the ear must be the final judge of the acoustic quality of a room.

REVERBERATION TIME—T_{60}

Reverberation time is arbitrarily defined as that time required for a sound to die away to one-thousandth of its initial sound pressure. As this corresponds to a drop in sound pressure level of 60 dB, it is convenient to abbreviate reverberation time as T_{60}. This definition is related to the characteristics of the human ear in that it represents very approximately a decay from a comfortable listening level to the threshold of audibility, say from 85 phons to 25 phons on Fig. 2-2. In fact, a stopwatch, a sound source (organ pipes were used historically), and a keen ear can be used to estimate reverberation times of 1 second or longer, found in the larger enclosures.

As sound decays in a room it follows what the mathematician calls an exponential curve. Plotted on a linear scale the exponential decays as shown in Fig. 9-1. Fig. 9-2 shows the same thing plotted as sound pressure level in dB, which is a logarithmic scale. On a logarithmic pressure scale, the exponential decay becomes a straight line which is more convenient to handle. On such a straight-line decay, the meaning of reverberation is as indicated in Fig. 9-2.

OPTIMUM REVERBERATION TIME

Too long a reverberation time leads to confusion of consecutive syllables in speech and a resulting loss of intelligibility. Too short a reverberation time results in a "dead" effect and loss of brilliance. This would lead one to suspect that there is some optimum reverberation time in between. Fig. 9-3 represents an attempt to summarize the acoustic "wisdom of the ages" in regard to optimum reverberation time as a function of room volume. As expected, the optimum reverberation time increases with the size of the listening room or studio. A small studio to be used for both music and speech can be designed around the shaded area of Fig. 9-3 which represents a modest compromise.

Fig. 9-3 appears to be clear-cut and definite. It is anything but that! It represents the author's subjective judgment on the subjective judgments of scores of authorities who do not agree among themselves on the optimum reverberation time. Because of this highly variable factor there is little sense in straining to achieve a T_{60} of 0.367 second when we are not even sure whether it should be 0.3 or 0.4. However, we can be assured that as far as reverberation times are concerned,

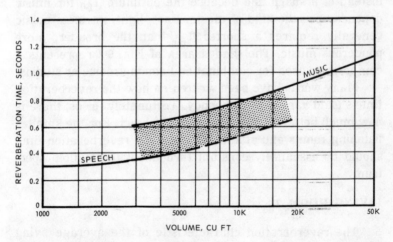

Fig. 9-3. Optimum reverberation time for small listening rooms and studios as a function of room volume. The shaded area between the speech and music curves may be considered a compromise region for studios to be used for both speech and music.

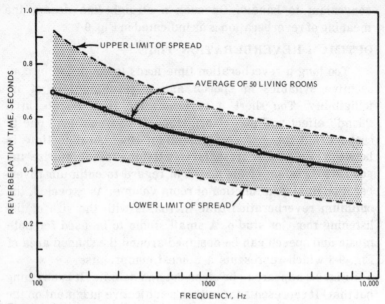

Fig. 9-4. Average reverberation time for 50 living rooms. (After Jackson and Leventhall.[18a])

following Fig. 9-3 will result in **reasonably** optimized and very usable conditions.

Optimum reverberation for speech lies on the single curve in Fig. 9-3. For music the optimum curve should be a fuzzy line instead of a sharp line because the optimum T_{60} for music depends upon the kind of music. Fast, light, intricate music generally requires a shorter T_{60} than the broader, more ponderous music. The shaded area of Fig. 9-3 represents a compromise area for combination listening rooms or studios.

Many words have been written on how the reverberation time should vary with frequency. Fortunately for us, there is now much better agreement on this subject. For the smaller listening rooms and studios at least, the reverberation time should be essentially constant throughout the audible spectrum.

LIVING ROOM T_{60}

The reverberation characteristic of the average living room interests the hi-fi enthusiast, the broadcaster, and the recording specialist. This living room is where the hi-fi buff listens to his recordings and the quality control monitoring

112

room of the broadcast and recording studio must have a reverberation time not too far from that of the living room in which the final product will be heard. Generally, such rooms should be "deader" than the living room which will add its own reverberation to that of the recording or broadcast studio.

Fig. 9-4 shows the average reverberation time of 50 British living rooms measured by Jackson and Leventhall using octave bands of noise[18a]. The average T_{60} decreases from 0.69 second at 125 Hz to 0.40 second at 8 kHz. This is considerably higher than earlier measurements on 16 living rooms made by BBC engineers[18b] which show a similar falling off with frequency. Presumably the living rooms measured by the BBC engineers were better furnished than those measured by Jackson & Leventhall.

The 50 living rooms were of varying sizes and degree of furnishing. The size varied from 880 to 2680 cu ft, averaging 1550 cu ft. From Fig. 9-3 we find an optimum T_{60} for speech for rooms of this size to be about 0.3 second. Only those living rooms near the lower limit approach this and in them we would expect to find much heavy carpet and overstuffed furniture. These T_{60} measurements tell us little or nothing about the possible presence of colorations. The BBC engineers checked for colorations and reported serious ones in a number of the living rooms studied.

TREBLE T_{60}

Having the reverberation time graph flat to 7 or 8 kHz is considered best for general listening and recording. Such a characteristic, however, may result in problems with brass instruments in small rooms. Trombones and cornets, for example, sound harsh and "brittle" with a flat curve while a droop in the treble makes these instruments sound better to the players and others in the room. On the other hand, the flat curve in the 7-10 kHz region makes woodwinds and strings sound better. A droop would be better for a voice rich in overtones. In summary, the drooping treble reverberation is safer for general conditions.

CALCULATION OF T_{60}—THE SABINE EQUATION

It is a relatively straightforward procedure to calculate the reverberation time of a listening room or a studio using the equation derived by W.C. Sabine, pioneer acoustician of Harvard University:

$$T_{60} = \frac{(0.05)(v)}{(S_T)(a_{ave})} \qquad (9\text{-}1)$$

where T_{60} = reverberation time, seconds

V = volume of room, cu ft

S_T = total surface area of room, sq ft

a_{ave} = average absorption coefficient

The quantity 0.05V involves the "mean free path" or the average distance a sound ray travels between reflections. Multiplying S_T by a_{ave} gives the total number of absorption units (sabins) in the room. As wall, ceiling, and floor may be covered with different materials, the absorption units of each material are determined separately and combined. The absorption coefficient, a, of a given material varies with frequency which makes it necessary to study the variation of T_{60} with frequency. The absorption coefficient is expressed as a decimal fraction of the sound energy absorbed; e.g., an a of 0.42 means that this fraction (or 42 percent) of the sound energy is absorbed, the rest reflected or transmitted as we saw in Chapter 1.

Equation 9-1 is based upon the assumption of perfectly diffuse sound conditions. This Sabine equation has the advantage of simplicity and ease of application but is accurate only for very live rooms which our listening rooms and small studios most certainly are not. However, to become familiar with the general technique of calculating reverberation time, it is expedient to use this simple equation and several oversimplified rooms and acoustic treatments just for practice.

Calculating reverberation time requires a bit of arithmetic, but nothing more. True, one's calculations may look a bit messy from time to time because of the standard procedure of carrying forward work on six frequencies simultaneously, but the basic steps for each frequency are quite simple. You should have no problem if you carefully follow every step of the examples. Don't worry about precision. A slide rule or an electronic or mechanical calculator would be timesavers but surely not required.

Example 1

As our first example let us take an untreated room (we don't care whether it is a listening room or a studio this time) 16x25 ft with a 10 ft ceiling. The floor is concrete; the walls and ceiling are smooth plaster. For the computation it is convenient to set up a tabular arrangement such as that in Table 9-1. The absorption coefficients for the surface materials, concrete and plaster, are obtained from Appendix I or from the literature.[31] Multiplying the area of concrete, 400 sq ft, by the absorption coefficient for concrete for 125 Hz (0.01) gives 4.0 sabins or absorption units at that frequency. Multiplying the 1220 sq ft of plaster by its coefficient (0.02) gives 24.4 sabins, making a total of 28.4 sabins or absorption units for 125 Hz. Dividing $0.05V=(0.05)(4,000 \text{ cu ft})=200$ by the total number of absorption units, 28.4 sabins, gives a T_{60} of 7.0 seconds for 125 Hz.

Following a similar procedure for other frequencies fills out Table 9-1 and yields data for plotting graph A in Fig. 9-5. This graph represents the "before treatment" condition of this room. Stepping into this untreated room, one would notice a serious blurring of speech as result of the exceedingly long reverberation time of 3.5 to 7.0 seconds. Careful listening would also reveal the fact that low-frequency sounds die away slower (have a longer T_{60}) than high-frequency sounds. We have now put numbers to a condition which has been recognized all along, that such a room is grossly unsuited for listening or recording. Perhaps we are beginning to see some reasons why this is so.

Example 2

Now let us cement some ordinary perforated acoustic tile (Graph A, Fig. 8-1 and Appendix I) to the ceiling only, without disturbing the plaster walls or the concrete floor. We recognize immediately that placing all the absorber on one surface flies in the face of everything we have learned earlier about distributing absorbing material, but for the sake of simplicity and a progressive example let's stick to it. The computations for this case are shown in Table 9-2 and the reverberation times are plotted as graph B in Fig. 9-5. With

Table 9-1. Room Conditions for Example 1

Size..... 16x25x10 ft
Treatment..none
Floor......concrete
Walls......plaster
Ceiling...plaster
Volume (16)(25)(10)=4000 cu ft

Material	S sq ft	125 Hz a	125 Hz Sa	250 Hz a	250 Hz Sa	500 Hz a	500 Hz Sa
Concrete	400	.01	4.0	.01	4.0	.015	6.0
Plaster	1220	.02	24.4	.02	24.4	.03	36.6
Total sabins			28.4		28.4		42.6
T_{60}, seconds			7.0		7.0		4.7

Material	1000 Hz a	1000 Hz Sa	2000 Hz a	2000 Hz Sa	4000 Hz a	4000 Hz Sa
Concrete	.02	8.0	.02	8.0	.02	8.0
Plaster	.04	48.8	.04	48.8	.03	36.6
Total sabins		56.8		56.8		44.6
T_{60}, seconds		3.5		3.5		4.5

S = area of material in sq ft

a = absorption coefficient for that material and for that frequency (See Appendix I)

Sa = S times a = absorption units in sabins

For 125 and 250 Hz:

$$T_{60} = \frac{(0.05)(V)}{\text{total sabins}} = \frac{(0.05)(4000)}{28.4} = \frac{200}{28.4} = 7.0 \text{ seconds}$$

only one surface treated, a dramatic reduction in T_{60} has been achieved. From Fig. 9-3 we can see that for a room of 4000 cu ft to be used as a combination music and speech, listening and recording room, the T_{60} should be roughly 0.5 second, holding

Table 9-2. Room Conditions for Example 2

Size........16x25x10 ft
Treatment...ceiling only treated
Floor.......concrete
Walls.......plaster
Ceiling.....Mtl. A, Fig. 8-1
Volume (16)(25)(10) = 4000 cu ft

Material	S sq ft	125 Hz a	125 Hz Sa	250 Hz a	250 Hz Sa	500 Hz a	500 Hz Sa
Concrete	400	.01	4.0	.01	4.0	.105	6.0
Plaster	820	.02	16.4	.02	16.4	.03	24.6
Acoustic Tile	400	.05	20.0	.20	80.0	.56	224.0
Total sabins			40.4		100.4		254.6
T_{60}, seconds			4.95		2.0		0.79

Material	1000 Hz a	1000 Hz Sa	2000 Hz a	2000 Hz Sa	4000 Hz a	4000 Hz Sa
Concrete	.02	8.0	.02	8.0	.02	8.0
Plaster	.04	32.8	.04	32.8	.03	24.6
Acoustic Tile	.95	380.0	.93	372.0	.74	296.0
Total sabins		420.8		412.8		328.6
T_{60}, seconds		0.48		0.49		0.61

S = area of material in sq ft

a = absorption coefficient for that material and for that frequency (see graph A Fig. 8-1)

Sa = S times a = absorption units in sabins

For 125 Hz:

$$T_{60} = \frac{(0.05)(V)}{\text{total sabins}} = \frac{200}{40.4} = 4.95 \text{ seconds}$$

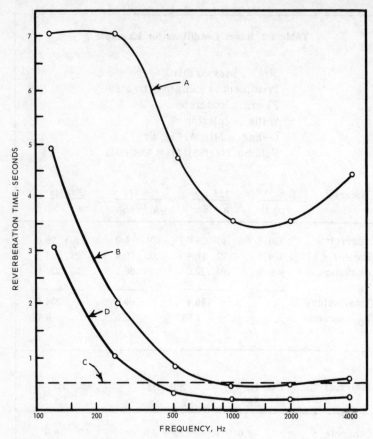

Fig. 9-5. Reverberation characteristics of a 16 x 25 x 10 ft room as an exercise in calculation.
(A) Example 1 before treatment
(B) Example 2, acoustic tile cemented to ceiling only
(C) Desired T_{60}
(D) Example 3, all walls and ceiling covered with acoustic tile. Conclusion: no amount of the common acoustic tile will yield the desired 0.5 second T_{60} throughout the band.

constant with frequency as shown by broken line C in Fig. 9-5. Graph B conforms nicely to this optimum T_{60} for frequencies above 800 Hz but very poorly for the low frequencies. This room would sound "boomy" as the low-frequency energy persists 8 or 10 times longer than high-frequency energy. The low-frequency region is still largely uncontrolled.

Example 3

Carrying this very unwise type of acoustic treatment one more absurd step, let us cover all walls with the same perforated acoustic tile as well as the ceiling. Without showing the detailed calculations, graph D of Fig. 9-5 results. Now the room is much too dead in the highs and still much too reverberant in the lows. Unfortunately, there are many small studios around the world treated just like this. In some they have compounded the problem by covering the floor with carpet which acts very much like more of the acoustic tile (compare graphs A and B in Fig. 8-1).

Identifying this perforated cellulose fiber tile as "acoustic tile" and their need as "acoustic treatment," there is apparently the feeling that the more of it used, the better the acoustics! This is completely fallacious. While the tile and the carpet are excellent acoustical materials when used intelligently, no amount of such materials will yield the balance required. What is needed is relatively more absorption at the low frequency end of the spectrum.

Example 4—Hi-Fi Living Room

In the first three examples we concentrated on one thing: how to compute reverberation time. We arbitrarily picked a 16x25x10 ft room with no consideration of the desirability of these proportions. Now let's try to make something of this room acoustically, first as a living room of a residence. It is our desire to make this room a good hi-fi listening room. Of course, the room serves many other family needs and we must not upset the general decor nor allow the costs to get out of line. We shall also assume that this room is already built and we are only making alterations.

Room proportions. First, let us see how these proportions check out as far as distribution of axial, tangential, and oblique modes is concerned. The room proportions (referring to ceiling height as unity) are 1.0 : 1.6 : 2.5. Check of Fig. 6-1 shows that this point is outside the area of acceptable ratios. What are we going to do about this? The little lady will probably object to shortening the room to bring it to one of the preferred ratios of Table 6-2. Can we live with it? How bad will it be and what is the nature of the faults?

Axial modes. Proper room proportions optimize the sound diffusion in the room, considering all modes. The axial modes, however, can cause the most trouble. Fortunately, they are the easiest to examine in detail.

The lowest axial mode will be that associated with the length of the room which resonates at 1130/(2)(25) or 23 Hz. The width will resonate at 1130/(2)(16) or 35 Hz and the height at 1130/(2)(10) or 56 Hz. In Table 9-3 these modes and their harmonics are spread out to a frequency of about 300 Hz. Combining all these in order of ascending frequency allows us to examine the spacing of the various modes. Let's cast out eye down the right-hand column of Table 9-3. We see modes near 70, 115, 140, 184, and 253 Hz which are separated from their neighbors by more than the 20 Hz specified in Chapter 5, but only a few hertz more. It is unlikely that these are spaced far enough to give trouble. We have a pile-up of two modes at 280 Hz, but they are well coupled to their neighbors and also very close to 300 Hz, the point above which isolated modes rarely prove troublesome.

We can breathe a sign of relief. Even though the room proportions are not optimized, the axial mode series seems quite acceptable. At least it is safe to proceed until a good listening test can be made. It is interesting to note that although these particular room proportions are off the "acceptable ratios" area of Fig. 6-1, they coincide exactly with the ratios recommended in earlier days by Volkmann and used in the constructions of many studios.

Room furnishings. As our room is the living room of a residence, we shall specify a wood floor and plaster walls and ceiling. The foldout sketch of the room (Fig. 9-6) includes only those furnishings having significant acoustical absorption. Actually, every small table, straight-backed chair, and picture on the wall contributes to diffusion of sound in the room, but only the softer items such as sofas, upholstered chairs, rugs, and drapes absorb much sound.

We can certainly hope that the hi-fi addict in the family would have something to say about the arrangement of the furniture. He would probably prefer having his loudspeakers aimed down the long dimension of the room as shown in Fig. 9-6 and Harry Olson tells us that they should be spaced about 0.7 of the room width, about 11 ft in this case. Placing the sofa about a room width away is within the good listening area. A couple of upholstered chairs as placed would help increase the seating capacity, but in somewhat less desirable positions. A

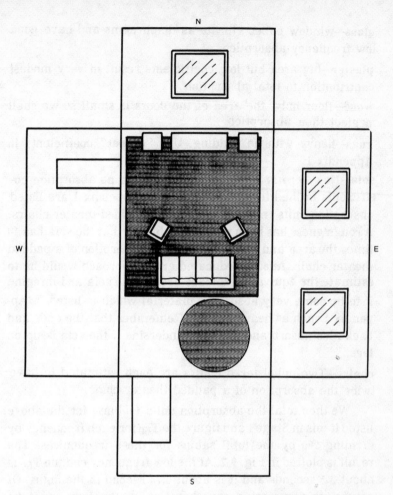

Fig. 9-6. Fold-out plan of hi-fi living room of Example 4 showing layout of furniture and rugs.

14x16 ft rug covers the north end of the room and a 7½ ft circular rug in the south end is a concession to a games area or a place to curl up with a good book.

T$_{60}$ calculation. Now comes the fun. What do we have to do with this living room to come out with an optimum reverberation time of about 0.5 second throughout the audible band? It is instructive to treat the problem in several stages, watching the effect on T$_{60}$ as new things are added. Table 9-4 shows the calculations for the six standard frequencies. Stage 1 includes the following:

glass—window panes vibrate as diaphragms and have good low frequency absorption.

plaster—big area but low coefficients result in very modest contribution to total absorption.

wood—floor only, the area of the doors is small so we shall neglect their absorption.

rugs—heavy with no padding. Use "carpet" coefficients in Appendix I.

sofa—Now here is a tough one. There are no absorption coefficients in the literature for such. In Appendix I are listed absorption units (not coefficients) for padded theater chairs. A rough guess has been made in Table 9-4 that the sofa has 10 times the area and hence 10 times the absorption of a padded theater chair. Another educated guess approach would be to estimate the equivalent surface area of the sofa and imagine it to be some very absorbent material which is listed in Appendix I such as heavy carpet. Remember that the ends and back also absorb and even the underside if the sofa is up on legs.

chairs—Two upholstered chairs are each estimated to have twice the absorption of a padded theater chair.

We then total the absorption units (sabins) for the above listed items in Stage 1 and figure the T_{60} for each frequency by dividing 200 by the total sabins for those frequenices. The result is plotted in Fig. 9-7. At the low frequency end the T_{60} is about 2.5 seconds and it is about 0.75 second in the highs. Of course, this excessive reverberation in the lows must be brought down.

In Stage 2 we add the drapes, having a height of 8 ft. The north window drapes are 9 ft wide and the east windows 7.5 ft giving a total of 192 sq ft of medium velour draped to half area (coefficients in Appendix I). This brings the T_{60} graph down some, but not enough (Fig. 9-7).

We have introduced rugs, drapes, and some upholstered furniture, and these materials leave us woefully short of low-frequency absorption.

In Stage 3 we add 700 sq ft of ⅛ in. paneling furred out from the plaster wall 3 inches with 1 in. of fiberglass absorbent in the cavity. This panel resonates at about 150 Hz (Fig. 8-3). If

Table 9-3. Axial Modes in Living Room of Example 4

Room Dimensions:

Length....25 ft
Width.....16 ft
Height....10 ft

AXIAL MODES - Hz

	L	W	H
f_0	23	35	56
$2f_0$	46	70	112
$3f_0$	69	105	168
$4f_0$	92	140	224
$5f_0$	115	175	280
$6f_0$	138	210	336
$7f_0$	161	245	
$8f_0$	184	280	
$9f_0$	207	315	
$10f_0$	230		
$11f_0$	253		
$12f_0$	276		
$13f_0$	299		

COMBINING L - W - H

AXIAL MODES	DIFFERENCE
23	12
35	11
46	10
56	13
69	19
70	22?
92	13
105	7
112	3
115	23?
138	2
140	21?
161	7
168	7
175	9
184	23?
207	3
210	14
224	14
230	15
245	8
253	23?
276	4
(280	0
280)	19
299	16
315	21
336	

Table 9-4. Room Conditions for Example 4

Material	S sq ft	125 Hz a	125 Hz Sa	250 Hz a	250 Hz Sa	500 Hz a	500 Hz Sa	1000 Hz a	1000 Hz Sa	2000 Hz a	2000 Hz Sa	4000 Hz a	4000 Hz Sa
Glass	65	.35	22.8	.25	16.2	.18	11.7	.12	7.8	.07	4.6	.04	2.6
Plaster	708	.02	14.2	.02	14.2	.03	21.2	.04	28.3	.04	28.3	.03	21.2
Wood	132	.15	19.8	.11	14.5	.10	13.2	.07	9.2	.06	7.9	.07	9.2
Carpet	268	.02	5.4	.06	16.1	.14	37.5	.37	99.0	.60	161.0	.65	174.0
Sofa			12.0		19.0		30.0		38.0		48.0		45.0
Chairs (2)			4.8		7.6		12.0		15.2		19.2		18.0
Stage 1 total sabins			79.0		87.6		125.6		197.5		269.0		270.0
Drapes	192	.07	13.4	.31	59.5	.49	94.0	.75	144.0	.70	134.0	.60	115.0
Stage 1 & 2 total sabins			92.4		147.1		219.6		341.5		403.0		385.0
Paneling	700	.55	384.0	.35	246.0	.11	77.0	.08	56.0	.07	49.0	.06	42.0
Stage 1, 2 & 3 total sabins			476.4		393.1		296.6		397.5		452.0		427.0
stage 1 T_{60}			2.53		2.28		1.59		1.01		0.75		0.74
Stage 2 T_{60}			2.16		1.36		0.91		0.59		0.50		0.52
Stage 3 T_{60}			0.42		0.51		0.67		0.50		0.44		0.47

Size......16x25x10 ft
Floor......wood
Walls......plaster
Ceiling......plaster
Volume (16)(25)(10) = 4000 cu ft

we cover all the walls with paneling, of course we should go back and subtract the absorption of the plaster on the walls which has been covered, but this is a minor refinement we won't bother with now. The T_{60} graph now is that shown for Stage 3 in Fig. 9-7. Not too bad! We have a slight T_{60} peak at 500 Hz and we're a bit over absorbed at 125 Hz.

Following the above procedure the reader can trace the various steps, making whatever changes are necessary to achieve the desired result.

There is a bit of a mystery here. Why did we have to add paneling to our walls to bring the low frequency T_{60} down? The average T_{60} of 50 British living rooms shown in Fig. 9-4 shows something of the same tendency to rise in the lows. We have no data on the individual houses but probably very few have paneled walls. The area of wood floor used in Stage 1 is that not covered by rugs. It is possible that a wood floor on wood joists is still reasonably effective as a low frequency absorber (0.15 at 125 Hz) even though covered by a rug. All the

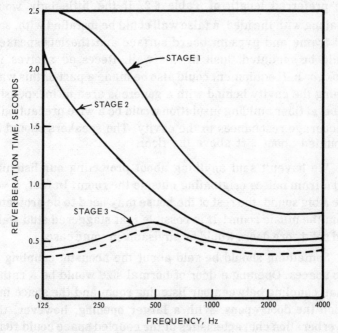

Fig. 9-7. Stage-by-stage changes in hi-fi living room reverberation.
(Stage 1) Including rugs, sofa, chairs
(Stage 2) Adding drapes
(Stage 3) Adding wood paneling

preceding calculations have assumed rigid massive walls, floor, and ceiling and that all the absorption is contributed by the surface treatment we apply. This neglects an important point, that with ordinary frame construction the room surfaces provide a certain amount of low-frequency absorption resulting from their diaphragm action. This is not included in any of our calculations but would be included in the Jackson-Leventhall measurements of homes in Fig. 9-4. All of this lends importance to actual measurements as described in Chapter 13.

Wood paneling has historically been associated with music rooms of Europe famous for their acoustics. The vibration of these panels seems to envelop the listener in the music. For this reason wood paneling is a natural and a logical material for achieving the low-frequency absorption which measurements indicate is needed and, with a thin veneer of an exotic wood, the paneling can be very rich and beautiful.

The length of our living room is about 1 ft 8 in. longer than the preferred length of Table 6-2. If the little lady would go along with the idea, a false wall could be installed with, say, 2x4 frame and gypsum board surface and the loudspeakers could be mounted flush in this wall. Recessed shelves for books or hi-fi equipment could also be made a part of this wall. Lining the cavity behind with a generous area of inexpensive mineral fiber building insulation would be a wise precaution to discourage resonances in the cavity. The speakers should be mounted about 5 ft above the floor.

We haven't said anything about protecting our listening room from noises originating outside the room. In fact, if you like a big sound, the rest of the house may need to be protected from the music room! It is possible that staggered stud walls and solid core doors would be advisable in some cases.

Something should be said about the acoustic coupling of two spaces. Opening a door of normal size would be a rather small coupling between our listening room and the space into which the door opens. With a larger opening, however, the reverberation characteristics of the coupled space could react on those of the listening room in a rather complicated way. The simplest solution, if such trouble develops, is to close the door!

We left our paneled walls flat and we know that some irregularities are needed to avoid flutter. Oil paintings are excellent for this and for a bit of low-frequency absorption of their own (stretched diaphragms with an air space behind). Other knick-knacks, gimcracks and bric-a-brac will help also. The wall opposite the loudspeakers may need some special attention. How about hanging a nice tapestry?

Example 5—Music Studio

To stretch our minds a bit in the calculation of reverberation time, let us take the same 16x25x10 ft room and make a recording studio out of it. Let us assume that the primary interest is in conventional music produced by small vocal and instrumental ensembles. Going back to Fig. 9-3 we would select for such use a T_{60} objective at the upper edge of the shaded area to favor music. For a volume of 4000 cu ft this would be near 0.6 second.

The proportions are not optimum for diffusion of sound according to Table 6-2. The C set of ratios would call for a room length of 23.3 ft instead of the existing 25 ft. We could shorten the room by 1.7 ft but, as we saw in Example 4, leaving it 25 ft long created no serious axial mode problems as far as we can determine, so we shall not shorten it.

In Chapter 8 we considered many different materials and structures capable of giving us the low frequency absorption we know is needed in a small studio. Because it is primarily a music studio, it is natural for us to think of paneling and because we are seeking good diffusion and brilliance for a recording studio, our thoughts go toward polycylindrical diffusers. It has been mentioned that polys are not much in favor today, but does anything as functional as polys really ever go out of style? In any event, we choose to go the poly route in this studio, which will also balance a later example utilizing modules.

Where does one start? First it is necessary to do some general thinking. Thoughts like this are in order:

(a) Let's see, a T_{60} of 0.6 second is desired. What absorption is required to get it? For a studio of 4000 cu ft volume, this will require how many sabins? From Equation 9-1 we compute this to be 333 sabins.

(b) Now, we will need 333 sabins at high frequencies and also that many at low frequencies. Polys can provide a large part of the low-frequency absorption and some needed diffusion as well. How many polys are needed to do the job?

Looking at the poly absorption graphs of Fig. 8-5 we see that in the 100-200 Hz region the absorption coefficient is 0.3 to 0.4; let's take an average of 0.35 as we will use polys of different sizes. To estimate the area of polys required we use the relationship:

$$\text{Absorption units (sabins)} = (\text{area})(\text{absorption coeff.})$$
$$\text{sabins} = S\,a$$
$$S = 333 \,/\, 0.35 = 950 \text{ sq ft}$$

Not all the low-frequency absorption will be from polys. The ceiling alone has 400 sq ft, so it would appear that we have wall space to accomodate 950 sq ft of polys.

(c) For high frequency absorption let's consider common acoustic tile such as used in Examples 2 and 3, **but we're using it carefully now to achieve a specific purpose!**

(d) Ideally we should have some of each acoustic material used applied to every surface of the studio. Practically, we can only approach this. For one thing, the only practical acoustic material for the floor is carpet and having the floor covered with carpet gives more absorption in the highs than we can stand.

A floor area of 400 sq ft and carpet absorption around 0.7 gives 280 sabins in the highs for the carpet alone and the polys have coefficients of 0.2 in the highs also. Even if we used only 500 sq ft of polys, they would contribute another 100 sabins. Carpet plus polys would give 280 + 100=380 sabins and we only want 333. So carpet is out. We'll settle for a throw rug directly under the microphone.

(e) We've theorized and nibbled around the edges about as much as we can, now we must dive into some preliminary, exploratory calculations. Basically, it is a cut-and-try process. The more experience one has, the fewer the exploratory calculations required. Taking the poly situation first, we know we should use polys of different chord lengths; let's pick three of the four shown in Fig. 8-5, poly A, poly B, and poly D. With a foldout sketch of the studio drawn to scale, you are ready to sketch possible configurations. In fact, it is handy to have a dozen photocopies of the foldout sketch to scribble on. Fig. 9-8

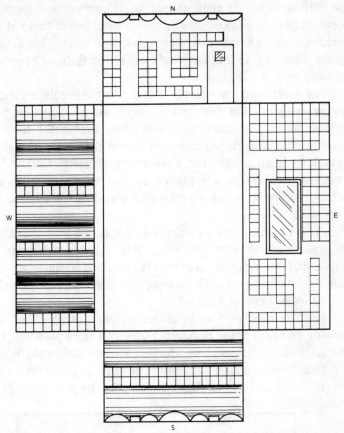

Fig. 9-8. Fold-out plan of music studio of Example 5. Note that the axes of the polys on the ceiling, the west wall, and the south wall are all mutually perpendicular.

is about the third or fourth such sketch and there were calculations for each.

On the north and south walls the cross section of the ceiling polys is shown. A big A poly in the center is flanked with a D poly on each side. Between the D polys and the outside B polys is space for fluorescent lighting fixtures running the length of the room. This well-polyd ceiling opposes the hard, reflective floor and should completely care for flutter in the vertical direction. It would have been nice to have some acoustic tile on the ceiling also, but this would have been a bit difficult to arrange.

On the west wall two A polys, three B polys and one D poly are arranged vertically which places them perpendicular to

129

the ceiling polys as they should be. Four vertical tiers of acoustic tile are worked into this west wall, two of them at the corners where absorbers are especially effective. This breaks up the west wall so that we would expect no flutter between it and the east wall.

The south wall has a single A and a single B poly arranged horizontally so that the axis of each set of polys is perpendicular to the other two sets. Two horizontal tiers of acoustic tile are placed between 4 and 6 ft from the floor (which is head height for a standing person). Good high-frequency absorbers are placed at this height on all walls except the one to which a performer would turn his back, the west wall.

(f) With the polys as selected in Fig. 9-8 and using the absorption coefficients plotted in Fig. 8-5 and tabulated in Appendix I, the absorption at the six standard frequencies was calculated (Table 9-5). The absorption attributed to the polys is the lower area of Fig. 9-9.

(g) The next step is to select an acoustic tile that will provide as much of the remaining absorption as possible. The tile offered by different companies varies somewhat in absorption characteristics. Many of these products were scanned with one eye on Fig. 9-9 with only the poly absorption

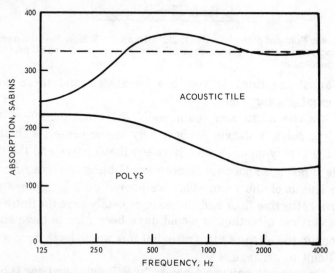

Fig. 9-9. The relative contributions of the polys and the acoustic tile to the overall absorption in the music studio of Example 5 and Fig. 9-8.

Table 9-5. Recording Studio—Example 5

Size.......16x25x10 ft
Floor.......vinyl tile
Walls.......plaster
Volume (16)(25)(10) = 4000 cu ft

Material	S sq ft	125 Hz a	125 Hz Sa	250 Hz a	250 Hz Sa	500 Hz a	500 Hz Sa	1000 Hz a	1000 Hz Sa	2,000 Hz a	2,000 Hz Sa	4000 Hz a	4000 Hz Sa
EMPTY													
Poly A	232	.41	95.0	.40	92.8	.33	76.5	.25	58.0	.20	46.4	.22	51.0
Poly B	271	.37	100.0	.35	95.0	.32	86.6	.28	75.8	.22	59.6	.22	59.6
Poly D	114	.25	28.5	.30	34.2	.33	37.6	.22	25.1	.20	22.8	.21	23.9
			223.5		222.0		200.7		158.9		128.0		134.5
FILLED													
Poly A	232	.45	104.3	.57	132.2	.38	88.0	.25	58.0	.20	46.4	.22	51.0
Poly B	271	.43	116.5	.55	149.0	.41	111.0	.28	75.9	.22	59.6	.22	59.6
Poly D	114	.30	34.2	.42	47.8	.35	39.8	.23	26.2	.19	21.6	.20	22.8
			255.0		329.0		238.8		160.1		127.6		133.4
Johns-Manville Spintone ½"	260	.09	23.4	.23	59.8	.62	161.0	.75	195.0	.77	200.0	.77	200.0
Total sabins EMPTY			246.9		281.8		361.7		353.9		328.8		334.5
T60 EMPTY, seconds			0.81		0.71		0.55		0.57		0.61		0.60
Total sabins FILLED			278.4		388.8		399.8		355.1		327.6		333.4
T60 FILLED, seconds			0.72		0.51		0.50		0.56		0.61		0.60

plotted. None of them do exactly what is needed but Johns-Manville Spintone, perforated 484 holes per sq ft and ½ in. thick cemented directly to the plaster surface, looked least bad.

A glance at Fig. 9-9 tells us that we need about 200 sabins at the high frequency end. This J-M Spintone offers an absorption coefficient of 0.77 in the highs and, following the same procedure as with the polys in (b) we find that we need 200 divided by 0.77 or 260 sq ft of Spintone. The upper part of Fig. 9-9 is plotted from the calculations of Table 9-5. This area of tile was then distributed around the studio in patches as shown in Fig. 9-8. There is nothing magical in this arrangement, only the knowledge that random patches contribute to diffusion of sound.

(h) Fig. 9-9 shows a deficiency of absorption in the lows. Fig. 8-5 shows that stuffing the polys full of mineral fiber increases absorption in the lows. Will this help us? Table 9-5 carries forward calculations for both empty and filled polys and the effect on the T_{60} graph is shown in Fig. 9-10. It would seem that all the trouble and expense of filling the polys, perhaps half of them, is questionable.

(i) Is the graph of Fig. 9-10, either for the "polys empty" or "polys filled" condition, close enough to our design objective of 0.6 second?

PLEASE SIT DOWN AND GRASP YOUR CHAIR FIRMLY. Here is some classified information. The calculations we have aren't all that good. We've plotted curves that were beautifully smooth and graceful. We've shown calculations to

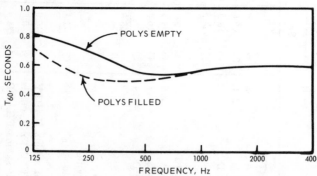

Fig. 9-10. The resulting reverberation characteristics of the music studio of Example 5 and Fig. 9-8.

three and four significant figures just to keep the columns neat. But, just between us girls, one T_{60} graph (Fig. 9-10) is about as good as the other (Fig. 9-9) **without some measurements** to tell us how to trim up our design. We shall see later that the Sabine reverberation equation we have used isn't very accurate for relatively dead rooms like ours. Therefore, there is no point to strsining further. Sounds recorded in this studio will almost certainly be very acceptable but we shall always have to be alert to the possibility of colorations due to some persistent axial mode.

THE EYRING REVERBERATION FORMULA

The Sabine formula of Eq. 9-1 has been used throughout this chapter because it is simpler than the Eyring formula which is more accurate for small listening rooms and recording studios. For those hardy souls who want the greater accuracy, Appendix II lays out step by step the method of using the Eyring approach.

THE "SOFT" STUDIO

Some studios are appearing with highly absorbent treatment which seems to negate all we have been learning about optimum reverberation time, sound diffusion, etc. These "dead" rooms are used for very special types of recording, primarily multitrack recording of rock music groups. The various instruments and vocalists are separated by distance and by reflective flats, and each is recorded on a separate track. This gives the producer complete freedom to achieve any balance he wishes in the mix as well as the opportunity of introducing any desired amount of special equalization, stereo effect, filter effects, or artificial reverberation.

In such studios the walls and ceilings may be covered with 4 inches of glass fiber batts, much as the conventional treatment of a motion picture sound stage. Surrounding each performer with plywood flats provides him with some sound return which enables him to hear his own instrument or voice. Without such "local" acoustics there is a tendency for a performer to feel he is not producing enough sound which he strains to correct.

TWO ROOMS COUPLED ELECTROACOUSTICALLY

What is the overall reverberant effect when sound picked up from a studio having on T_{60} is reproduced in a listening room having a different T_{60}? Does the listening room reverberation affect what is heard The answer is definitely, yes. This problem has been analyzed mathematically by Mankovsky.[17] In brief, the sound in the listening room is affected by the reverberation of both the studio and the listening room as follows:

(a) The combined T_{60} is greater than either alone,

(b) The combined T_{60} is nearer the longer T_{60} of the two rooms,

(c) The combined decay departs somewhat from the straight line of Fig. 9-2,

(d) If one room has a very short T_{60}, the combined T_{60} will be very close to the longer one,

(e) If the T_{60} of each of the two rooms alone is the same, the combined T_{60} is 20.8 percent longer than one of them,

(f) The character and quality of the sound field transmitted by a stereo system conforms more closely to the mathematical assumptions of the above than does a monaural system, and

(g) Items (a) to (e) can be applied to the case of a studio linked to an echo chamber as well as a studio linked to a listening room.

ACOUSTIC DESIGN OF A STUDIO

10

Tracing each step in the acoustic design of a specific suite of general purpose studios is the plan of this chapter. By doing this we can get the feel of how a job is approached and for the inevitable compromises involved. There will be similarities and differences to the five examples of Chapter 9.

This suite of studios is to be used for the recording of programs for later use by radio stations. The general specifications are laid down by the client—how many studios, their general size and arrangement, and their relation to other functions such as office, tape library, and various supporting services.

An analysis is first made of automobile traffic and aircraft noise levels at the proposed site. This information is helpful in wall and roof specifications.

The general layout of the part of the building housing the studios is agreed upon after a series of consultations. See Fig. 10-1. One control room serves both the main studio and a small speech studio. The studios are entered through their respective sound locks which make possible entering a studio while it is in use. These sound locks were considered to be necessary because of high noise levels in the reception area and service areas east of the control room and speech studio.

MAIN STUDIO

The dimensions of the main studio were established at 30 ft 0 inches by 20 ft 8 inches with a ceiling height of 12 ft 11 inches. These dimensions yield ratios corresponding to ratio set C of Table 6-2. The volume turns out to be 7782 cu ft.

Just to verify the axial mode situation, Table 10-1 has been prepared. We note that the larger dimensions of this studio, as compared to our five examples of the previous chapter, yield

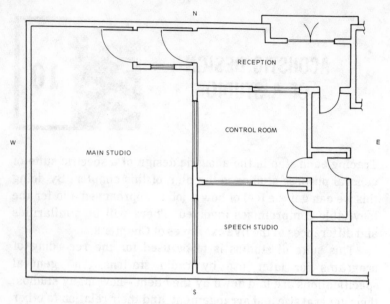

Fig. 10-1. Floor plan for general purpose studio.

axial modes of lower frequencies which, in turn, mean closer spacings. An examination of the difference column in Table 10-1 reveals that the greatest spacing of axial modes below 300 Hz is 19 Hz which is less than our 20 Hz criterion. Therefore, the main studio would appear to be safe from axial mode colorations.

Reverberation Time

Fig. 9-3 tells us that, for a 7782 cu ft studio, the optimum T_{60} for music is about 0.75 second, for speech about 0.53 second. As this studio is to be used for both music and speech, we select a design objective of about 0.64 second with assurance of reasonable effectiveness for both. There is the possibility of using variable acoustic elements in the form of swinging panels, etc., to optimize for speech or music, according to the present use of the studio. This approach was rejected as unnecessary for the uncritical work contemplated.

Floor Treatment

A decision on floor treatment is a good place to start because of the overpowering influence of 593 sq ft of anything

Table 10-1. Axial Modes in Main Studio

Room dimensions:

Length.....30 ft 0 in (30.0 ft)
Width......20 ft 8 in (20.66 ft)
Height.....12 ft 11 in (12.9 ft)

COMBINING L-W-H

AXIAL MODES - Hz				COMBINING AXIAL MODES	L-W-H DIFFERENCES
	L	W	H	19	
					8
				27	
					11
f_0	19	27	44	38	
					6
$2f_0$	38	54	88	44	
					10
$3f_0$	57	81	132	54	
					3
$4f_0$	76	108	176	57	
					19
$5f_0$	95	135	220	76	
					5
$6f_0$	114	162	264	81	
					7
$7f_0$	133	189	308	88	
					7
$8f_0$	152	216		95	
					13
$9f_0$	171	243		108	
					6
$10f_0$	190	270		114	
					18
$11f_0$	209	297		132	
					1
$12f_0$	228			133	
					2
$13f_0$	247			135	
					17
$14f_0$	266			152	
					10
$15f_0$	285			162	
					9
$16f_0$	304			171	
					5
				176	
					13
				189	
					1
				190	
					19
				209	
					17
				216	
					4
				220	
					8
				228	
					15
				243	
					4
				247	
					17
				264	
					2
				266	
					4
				270	
					15
				285	
					12
				287	
					7
				304	
					4
				308	

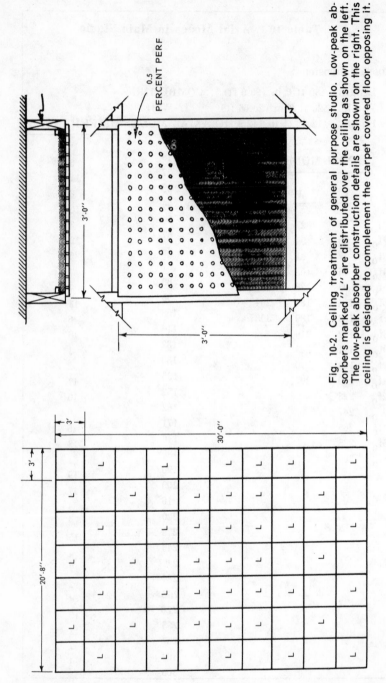

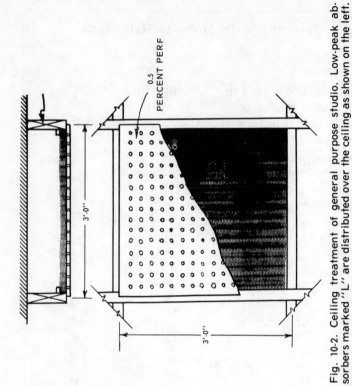

Fig. 10-2. Ceiling treatment of general purpose studio. Low-peak absorbers marked "L" are distributed over the ceiling as shown on the left. The low-peak absorber construction details are shown on the right. This ceiling is designed to complement the carpet covered floor opposing it.

0.5 PERCENT PERF

in a studio of this size. Certainly, designing around a vinyl or asphalt tile floor with the idea that it can be changed later will not work. In the present case the client expresses a strong desire for wall-to-wall carpet. Carpet on sponge rubber underlay has been specified.

Ceiling Treatment

The ceiling is concrete. Opposing the carpeted floor with concrete would be very bad in the low frequencies at which the carpet is a poor absorber. The BBC uses a vivid term for what is needed, "anticarpet." That is, a material or structure which tends to complement the lopsided absorption of the carpet by being lopsided in the opposite way. The final decision on type, amount, and distribution of this low-peak absorber for the ceiling is shown in Fig. 10-2.

The entire ceiling is covered with parallel 2x8s mounted edge-on to the ceiling at 3-foot centers. These can be either parallel to the long or the short axis of the room. The installation of short pieces of 2x8 at right angles to the parallel 2x8s form squares 3x3 ft on centers. Each square marked "L" has built into it the low-peak absorber described on the right side of Fig. 10-2. Those squares without an "L" are left open with concrete ceiling showing.

The covers of the low-peak absorbers are perforated 0.5 percent, giving the absorption characteristics as shown in Fig. 8-11. It is unlikely that 3/16 or 1/4 in. plywood or hardboard can be found already perforated this amount. It can be done by hand by stacking the panels and using a pattern with guide holes drilled in it. Holes 11/64 in. diameter spaced 2-1/4 in. on centers work out about right. Other hole sizes and spacings may be used if the perforation area is kept between 0.4 and 0.6 percent of the total area. The fiberglass board is semirigid, requiring a minimum of support to hold it against the back of the perforated panel.

The air space behind the perforated covers and the fiberglass board should be airtight except through the face. The 2x8s should either be set in mastic compound against the ceiling or a fillet of caulking compound should be run around the inside of each 3x3 ft unit enclosed.

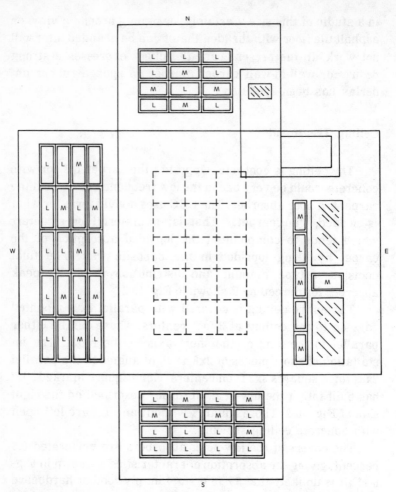

Fig. 10-3. Arrangement of modules for acoustic treatment of walls of main studio of Fig. 10-1. "L" indicates low-peak modules similar to those on the ceiling while "M" indicates mid-peak modules. The broken lines indicate a suspended overhead feature to provide for illumination fixtures, and air conditioning grills and to hide the practical-but-not-so-beautiful ceiling.

Wall Treatment

The foldout drawing of the main studio, Fig. 10-3, shows the location of 54 acoustic modules, each of 2x4 ft size. They are of two types, low-peak absorbers marked "L" having 0.5 percent perforated covers such as are on the ceiling, and mid-peak absorbers marked "M" having 5 percent perforated covers. The construction of these modules is shown in Fig. 8-12 and their absorption characteristics are listed in Appendix I and plotted in Fig. 8-11.

The 33 low-peak modules (L) and 21 mid-peak modules (M) are distributed more or less randomly between the four wall surfaces. This completes the acoustic treatment of the main studio.

Fig. 10-4 shows graphically the absorption contribution of the three main acoustic elements, the carpet and the low- and high-peak modules. The low-peak modules nicely complement the carpet and the mid-peak modules add more needed absorption and straighten up the total absorption line a bit. The absorption of plaster and glass is so small that it has been neglected.

The reverberation times calculated in Table 10-2 are by the simpler, Sabine method. These are plotted in Fig. 10-5 as a broken line. The solid line of the same figure is the reverberation time calculated by the more rigorous Eyring formula, calculations not shown (refer to Appendix II). Just to see how close to the Eyring results the Sabine T_{60} minus the 0.027 V divided by S correction would come, this computation was also carried out. The correction turns out to be 0.082 and subtracting this amount from the Sabine values (broken line of Fig. 10-5) comes very close to the Eyring figures. This suggests that, under these circumstances, the labor of computing the Eyring T_{60}s was unnecessary.

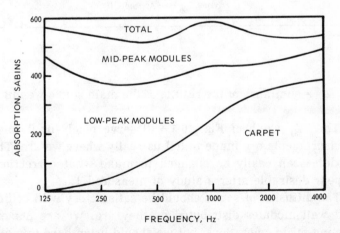

Fig. 10-4. The absorption contribution of the low- and mid-peak absorbers and the carpet in the main studio of Fig. 10-1.

Table 10-2. Main Studio

Size.....20.66x30.0x12.92 ft
Floor....Carpet on sponge rubber underlay
Walls....per Fig. 10-3
Volume...7782 cu ft
Surface area....2550 sq ft

Material	S sq ft	125 Hz a Sa	250 Hz a Sa	500 Hz a Sa	1000 Hz a Sa	2000 Hz a Sa	4000 Hz a Sa
Carpet	593	.00 0	.05 30.	.20 119.	.40 247.0	.60 355.	.65 385.
Ceiling (Low-Peak)	378	.74 280.	.53 200.	.40 151.	.30 113.	.14 52.8	.16 60.5
Wall (Low-Peak)	264	.74 195.	.53 140.	.40 105.5	.30 79.1	.14 36.9	.16 42.2
Wall (Mid-Peak)	168	.60 101	.98 164.6	.82 138.	.90 151.3	.49 82.4	.30 50.4
Total sabins...		576.0	534.6	513.5	590.4	527.1	538.1
Sabine T_{60},seconds		0.675	0.728	0.756	0.658	0.736	0.722
Eyring T_{60},seconds		0.595	0.645	0.678	0.578	0.658	0.645

$$\text{Sabine } T_{60} = \frac{(0.05)(V)}{\text{Total sabins}} = \frac{(0.05)(7782)}{\text{Total sabins}} = \frac{389}{\text{Total sabins}}$$

Eyring T_{60} (see Appendix II)

As a summary of the results of the main studio we might say:

1. The T_{60} graph of Fig. 10-5 will serve nicely until some measurements are made to tell us really where we are. The modules can easily be changed to make what corrections appear desirable after a study of measured T_{60}.

2. The diffusion of sound should be satisfactory with ceiling and wall modules distributed as they are. We are not expecting mode colorations, but we should listen hard to make sure there are none.

3. Flutter should be absent because opposing parallel surfaces are treated with absorbing materials.

4. The appearance of the ceiling is nothing to rave about so we shall do something about that as the next step.

Overhead Feature

It is suggested that an overhead feature about 12x20 ft be suspended from the ceiling with about 9 ft 6 inches clearance above the floor. This feature should be an optical barrier, but not an acoustical barrier. In other words, it should not divide the studio into an acoustic upstairs and downstairs. There are available plastic sheets composed of vertical cellular elements. Perhaps these would serve to shield the upper part from the eye and be quite open to sound waves. Wooden slats could do the same. A good architect could design a feature which would enhance the appearance greatly. This suspended feature would also support the lighting fixtures and air conditioning grills. Everything should be painted flat black from the top of the wall modules upwards and care should be exercised **not** to fill the holes in the ceiling cover boards with paint. If the illumination fixtures throw their light downward, the air conditioning ducts and ceiling units should not be visible.

SPEECH STUDIO

As we take up the task of treating very small rooms such as the control room and speech studio, we are faced with the

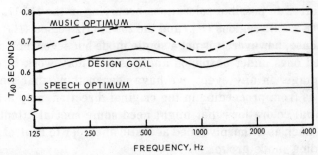

Fig. 10-5. The calculated reverberation time of the main studio of Fig. 10-1. Broken line: Sabine formula; solid curve: Eyring formula (the latter is the more accurate for small rooms).

acute problem of avoiding, or at least minimizing, axial mode colorations. A high percentage of the wall area is glass and doors, leaving only a limited space for absorbers designed to provide needed low frequency absorption. In such rooms the "suspended" or "lay-in" or "drop" ceiling offers the possibility of good absorption throughout the 125-4000 Hz band **if the lay-in material is carefully selected.** This type of ceiling treatment has been selected for both the control room and speech studio. This drop ceiling is placed 16 inches below the plastered concrete ceiling which is described in the American literature as Mounting No. 7.[31]

Axial Modes

Using Johns-Manville Acousti-Shell TF flat of ⅛ in. thickness, absorption coefficients averaging 0.75 are achieved. This material acts as a diaphragm and, with the 16 inch cavity, is a resonant system. If 75 percent of the sound falling on the drop ceiling is absorbed, what is the true acoustic height of the room, 9 ft 10 inches or 8 ft 6 inches? This may be considered a relatively minor point, but it is involved in the checking of axial modes. The 8 ft 6 in. ceiling height gives precisely the preferred ratio set B in Table 6-2. The 9 ft 10 in. height comes close to ratio set A in the same table. Just to be on the safe side, let's consider both!

Table 10-3 shows the axial mode tabulation for both heights H_1 (9 ft 10 in.) and H_2 (8 ft 6 in.). Glancing at the combined mode and difference columns for H_1 we see that there are five differences greater than 20 Hz below 300 Hz. For the H_2 combination we see that there are also five. We should be alerted to possible colorations occurring in the 156-172 Hz and 208-228 Hz regions (H_1) and the 258-264 Hz region (H_2). In each case, however, it is not a single mode but several closely spaced ones under suspicion and this fact may work to our advantage. In any event, we have seen nothing that would keep us from proceeding in the original direction, only a few potential colorations that might need some special attention later, such as a sharply tuned acoustic absorber to control the offending mode group.

The distribution of the various acoustic elements in the speech studio are shown in Fig. 10-6. The drop ceiling of J-M

Table 10-3. Axial Modes in Speech Studio and Control Room

Room dimensions:

        Length....13 ft 1 in. (13.08 ft)
        Width.....10 ft 11 in. (10.92 ft)
        Height....(H_1) 9 ft 10 in. (9.83 ft)
        Height....(H_2) 8 ft 6 in. (8.50 ft)

AXIAL MODES - Hz				COMBINING L - W - H_1		COMBINING L - W - H_2	
L	W	(H_1)	(H_2)	MODES	DIFF.	MODES	DIFF.
f_0 43	52	57	66	43		43	
$2f_0$ 86	104	114	132		9		9
$3f_0$ 129	156	171	198	52		52	
$4f_0$ 172	208	228	264		5		14
$5f_0$ 215	260	285	330	57		66	
$6f_0$ 258	312	342			29 ?		20
$7f_0$ 301				86		86	
					18		18
				104		104	
					10		25 ?
				114		129	
					15		3
				129		132	
					27 ?		24 ?
				156		156	
					15		16
				171		172	
					1		26 ?
				172		198	
					36 ?		10
				208		208	
					7		7
				215		215	
					13		43 ?
				228		258	
					30 ?		2
				258		260	
					2		4
				260		264	
					25 ?		37 ?
				285		301	
					16		11
				301		312	
					11		18
				312		330	
					30?		
				342			

Acousti-Shell provides approximately half the required absorption throughout the 125-4000 Hz band as shown in Table 10-4 and in Fig. 10-7. The rug must be limited to 8x8 to avoid overabsorption in the highs. Most of the remaining needed absorption is contributed by two types of wall-mounted structures. The three low-peak modules are identical to those of the walls and ceiling of the main studio (0.5 percent graph in

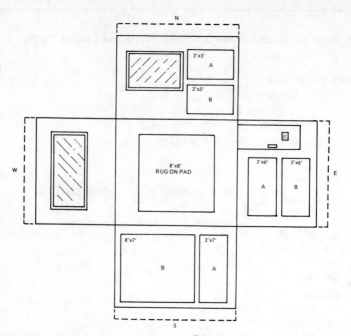

Fig. 10-6. Fold-out plan of speech studio of Fig. 10-1 showing wall elements. The "L" low-peak elements are the same type as the wall and ceiling L-units in the main studio. The mid-peak absorbers are somewhat different, fabricated as shown in Fig. 10-8.

Fig. 8-11 and the structure of Fig. 8-12) except for module shape and area. The mid-peak modules in the speech studio are somewhat different from those in the main studio. They are fabricated around 2x2-inch frames with 0.5 percent perforated cover with the space filled with 2 inches of the same 5 to 8 lb per cu ft density fiberglass as shown in Figs. 8-12 and 10-8. The total contribution of the low-peak and mid-peak modules, the drop ceiling, and the rug are almost constant with frequency as shown in Fig. 10-7.

Reverberation Time

A 1400 cu ft studio intended primarily for speech use should have a T_{60} of 0.3 second (Fig. 9-3). The computations of Table 10-4 show that the Sabine T_{60} is almost 10 percent higher than this. This was intentional, knowing that the Eyring figures would be lower, a mite too low, as it developed, but not worth correcting until measurements are made. Here again, corrrecting the Sabine T_{60} by subtracting the 0.027 V divided

146

by S factor of 0.05 gives almost exactly the same results as the more laborious Eyring computation of Appendix II. The final reverberation time is nicely flat with frequency (Fig. 10-9).

Flutter

We cannot consider the design of the speech studio finished until we have checked for possible flutter problems. Perforated plywood or hardboard facings, even though covered with grill cloth, can produce flutter because of the significant reflection from their surfaces. Referring to Fig. 10-6, if the glass in the west window is inclined from the vertical, coherence in the east-west direction should be quite well destroyed. The inner surface of the door could be padded with a plastic-impregnated fabric over 1'' foam rubber quilted with upholstery tacks. In the north-south direction the glass should again be inclined to the vertical. The two modules on the north wall working against the single one on the south wall do have parallel surfaces. If trouble should develop, modules could be

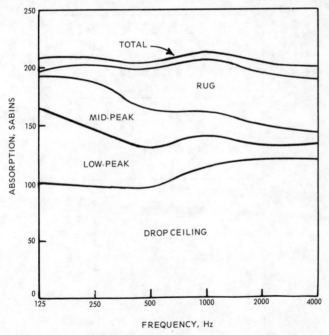

Fig. 10-7. The absorption contributions of the four principal acoustic elements of the speech studio.

147

Table 10-4. Speech Studio

Size....10.91 x 13.09 x 9.83 ft
Floor...Heavy 8x8 rug on vinyl tile floor
Walls....per Fig. 10-6
Volume...1400 cu ft (to plaster ceiling)
Surface area....758 sq ft (to plaster ceiling)

Material	S sq ft	125 Hz a	125 Hz Sa	250 Hz a	250 Hz Sa	500 Hz a	500 Hz Sa	1000 Hz a	1000 Hz Sa	2000 Hz a	2000 Hz Sa	4000 Hz a	4000 Hz Sa
Glass	56	.18	10.1	.06	3.4	.04	2.2	.03	1.7	.02	1.1	.02	1.1
Plaster	208	.013	2.7	.015	3.1	.02	4.1	.03	6.2	.04	8.3	.05	10.4
Drop Ceiling*	143	.70	100.0	.69	98.5	.66	94.0	.80	114.5	.84	120.0	.83	119.0
Rug	64	.08	5.1	.24	15.4	.57	36.4	.69	44.2	.71	45.5	.73	46.7
Mid-peak (shallow)	54	.48	25.9	.78	42.1	.60	32.4	.38	20.5	.32	17.3	.16	8.7
Low-peak	89	.74	65.8	.53	47.2	.40	35.6	.30	26.6	.14	12.4	.16	14.2
Total sabins......			209.6		209.7		204.7		213.7		204.6		200.1
Sabine T_{60}, seconds			0.334		0.334		0.341		0.328		0.342		0.350
Eyring T_{60}, seconds			0.284		0.284		0.292		0.278		0.292		0.300

Sabine $T_{60} = \dfrac{(0.05)(V)}{\text{Total sabins}} = \dfrac{(0.05)(1400)}{\text{Total sabins}} = \dfrac{70}{\text{Total sabins}}$

Eyring (see Appendix II)

*Johns-Manville Acousti-Shell TF flat 1/8 inch

Table 9-5. Recording Studio

Size....10.92 x 13.08 x 9.83 ft
Floor...Indoor-outdoor carpet
Walls...Per Fig. 10-10
Volume..1400 cu ft (to plaster ceiling)
Surface area..758 sq ft (to plaster ceiling)

Material	S sq ft	125 Hz a	125 Hz Sa	250 Hz a	250 Hz Sa	500 Hz a	500 Hz Sa	1000 Hz a	1000 Hz Sa	2000 Hz a	2000 Hz Sa	4000 Hz a	4000 Hz Sa
Glass	80	.18	14.4	.06	4.8	.04	3.2	.03	2.4	.02	1.6	.02	1.6
Plaster	263	.013	3.4	.015	3.9	.02	5.3	.03	7.9	.04	10.5	.05	13.1
Drop Ceiling	143	.70	100.0	.69	98.5	.66	94.0	.80	114.5	.84	120.0	.83	119.0
Carpet	143	.01	1.4	.05	7.2	.10	14.3	.20	28.6	.45	64.3	.65	93.0
Low-peak	64	.48	30.7	.78	49.9	.60	38.4	.38	24.3	.32	20.5	.16	10.2
Total sabins......			149.9		164.3		155.2		177.7		216.9		236.9
Sabine T$_{60}$, seconds			0.466		0.426		0.451		0.394		0.323		0.295
Eyring T$_{60}$, seconds			0.416		0.376		0.401		0.344		0.273		0.245

$$\text{Sabine } T_{60} = \frac{(0.05)(V)}{\text{Total sabins}} = \frac{(0.05)(1400)}{\text{Total sabins}} = \frac{70}{\text{Total sabins}}$$

Eyring (see Appendix II)

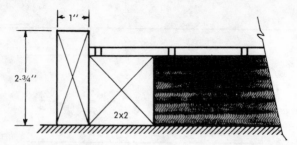

Fig. 10-8. The construction of the mid-peak modules of the speech studio.

inclined to the vertical. If this were necessary, a back should be put on the mid-peak modules. The low-peak module frame could be inclined to give an **average** depth of 8 inches with the cavity sealed in the usual way (Fig. 8-12).

CONTROL ROOM

Tests have indicated that the most consistent listening in a control room results if the reverberation time is kept below 0.4 second. In fact, Gilford [18b] reports that BBC control rooms are designed to have a T_{60} of 0.4 second up to 1000 Hz, decreasing to 0.3 second above this frequency. Let us take this as our design goal in the control room under consideration.

Control Room Treatment

Fig. 10-10 shows the end result of considerable cutting and trying. Table 10-5 shows the calculations and Fig. 10-11 the graphical presentation of the same information. Again the drop ceiling with J-M Acousti-Shell and a cavity depth of 16 inches dominates the absorption picture across the band, just

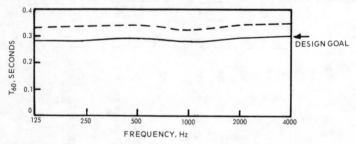

Fig. 10-9. Calculated reverberation characteristics of the speech studio. Broken line: Sabine formula; solid curve: Eyring formula.

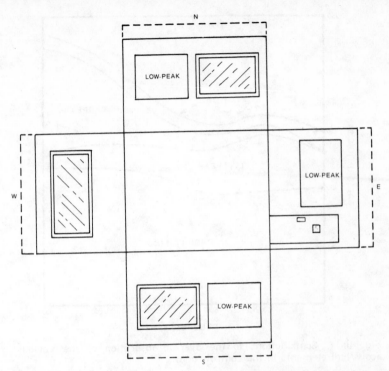

Fig. 10-10. Fold-out plan of control room of general purpose studios of Fig. 10-1.

like the speech studio. Three double glass windows grace this control room, one for the people in the reception room to gaze through. Three low-peak panels of the Fig. 10-8 design are used. The carpet selected is the inexpensive, felt-like "indoor-outdoor" type which is less absorbent than heavy carpet on a pad and yet very durable and practical under heavy service.[40]

Reverberation Time

Fig. 10-12 gives the Eyring T_{60} results for the control room from the calculations of Table 10-5. The broken line represents the design goal which is not very closely met above 1000 Hz. Looking at Fig. 10-12 along with Fig. 10-11, it is obvious that T_{60} could be raised a bit simply by reducing the area of the indoor-outdoor carpet. The control room will be full of equipment and some built-in work is possible under the windows. It would seem unwise to waste time on such trimming until the console, recorders, and cabinets are in place and actual reverberation measurements have been made.

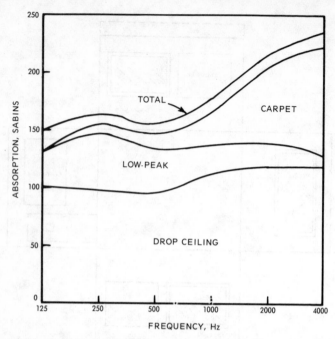

Fig. 10-11. Contributions to the overall absorption of the various acoustical elements in the control room.

GENERAL DESIGN FACTORS

The floor in both sound locks should be carpeted to reduce foot noise. The walls and ceilings of the sound locks should be made as absorptive as possible consistent with good scuff resistance. There is a treatment for such tight areas subject to

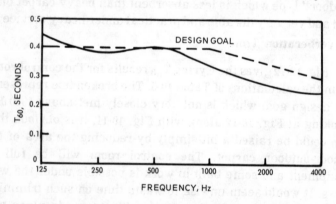

Fig. 10-12. Calculated reverberation characteristics of the control room. The broken line represents the design goal.

much abrasion which is almost traditional, probably because it has proved so durable and effective. The covers are Johns-Manville Transite of 3/16'' asbestos board perforated with 550 3/16 inch holes per sq ft. Mounting this on 2x2-inch furring strips with the space behind filled with a mineral fiber batt makes an effective lining for sound traps.

After all that has been said about walls, windows, and doors in Chapter 7 and their part in the control of interfering noise, there is no reason to dwell further on these subjects in connection with the design of this suite of studios.

11 | ADJUSTABLE ACOUSTICS

If a space is used for only one purpose, it can be treated acoustically with some precision. A multipurpose studio, on the other hand, carries with it compromises, the magnitude of which can be roughly estimated by the difference between the optimum reverberation times for music and speech of Fig. 9-3. However, we remember that Fig. 9-3 is a subjective appraisal of the result of many subjective listening tests and cannot be taken as the whole truth.

There is general agreement that the T_{60} of small studios should be essentially uniform thoughout the audible range but a voice rich in overtones might be better served by a T_{60} curve that droops in the highs. If the reverberation time remains flat above 7 kHz, cornets and trombones tend to sound harsh and even though the reproducing, recording, or broadcast channel may not be too effective in the 7-10 kHz region, the effect can be very distracting to other performers, especially in the smaller studios. A compromise is almost always necessary and experience indicates that compromising in the too-dead direction is preferable.

We tend to think of acoustic materials as something to nail or cement to a surface, but let us look at this more carefully. It is possible to pull a drape having one type of absorption characteristic in front of a surface having quite a different characteristic. Hinged panels, rotating elements, or simple portable elements can alter the absorption and hence the reverberation characteristics of a room. With variable elements it is possible to change a flat reverberation characteristic from one value to another. The actual shape of the reverberation-frequency graph can be changed with or without a change in average reverberation time.

In any given studio situation, the decision must be made as to how much second-order fussing is justified. Perhaps in most cases a single compromise reverberation characteristic would be considered adequate. However, it is well for us to be aware of the techniques for varying the acoustic quality of an enclosure, many of which are quite straightforward and inexpensive.

DRAPERIES

As radio broadcasting developed in the 1920s, draperies on the wall and carpets on the floor were almost universally used to "deaden" studios. During this time there was remarkable progress in the science of acoustics. It became more and more apparent that the old radio studio treatment was quite unbalanced, absorbing middle- and high-frequency energy but providing little absorption at the lower frequencies. As proprietary acoustic materials became available, hard floors became common and drapes all but disappeared from studio walls.

A decade or two later the acoustical engineers, interested in adjusting the acoustical environment of the studio to the job to be done, turned with renewed interest to draperies. A good example of this early return to draperies was illustrated in the rebuilding of the old studio 3A of the National Broadcasting Company of New York City in 1946. This studio was redesigned for optimum conditions for records for home use, and transcriptions for broadcast purposes. The acoustical criteria for these two jobs differ largely as to the reverberation-frequency characteristic. By the use of drapes and hinged panels (to be considered later), the reverberation time was made adjustable over more than a two-to-one range. The heavy drapes were lined and interlined and were hung some distance from the wall to make them more absorbent at the lower frequencies. When the drapes were withdrawn, polycylindrical elements having a plaster surface were exposed.[41] (Plywood was in critical supply in 1945-46.)

If due regard is given to the absorption characteristics of draperies, there is no reason other than cost why they should not be used. The effect of the fullness of the drape must be considered (Fig. 8-2). The acoustical effect of an adjustable element using drapes can thus be varied from that of the drape itself when closed (Fig. 11-1) to that of the material behind

Fig. 11-1. The reverberation of a room may be varied by pulling absorptive drapes in front of reflective areas.

when the drapes are withdrawn into the slot provided. The wall treatment behind the drape could be anything fron hard plaster for minimum sound absorption to resonant structures having maximum absorption in the low-frequency region, more or less complementing the effect of the drape itself. Acoustically, there would be little point to retracting a drape to reveal material having similar acoustical properties.

PORTABLE PANELS

Portable absorbent panels offer a certain amount of flexibility in adjusting listening room or studio acoustics. The simplicity of such an arrangement is illustrated in Figs. 11-2 A and B. In this example[42] a perforated hardboard facing, a mineral fiber layer, and an air cavity constitute a low frequency resonator. Hanging such units on the wall adds low-frequency absorption, acoustically removes some of the highly reflective wall surface, and contributes somewhat to sound diffusion by its shape. There is some compromising of the effectiveness of the panels as low-frequency resonators in that the units hang loosely from the mounting strip. "Leakage" coupling between the cavity and the room would tend to destroy the resonant effect. Panels may be removed to obtain a "live" effect for instrumental music recording, for example, or introduced for voice recording. They may be easily positioned to "kill" troublesome reflections.

Free standing "acoustic flats" are useful studio accessories. A typical flat consists of a frame of 1x4 lumber inside of which is supported a low density (e.g., 4 lb per cu ft) mineral wool blanket faced on both sides with a sound-transparent fabric such as muslin or glass fiber cloth to protect the soft surface. Arranging a few such flats strategically can give a certain amount of local control of acoustics.

ROTATING ELEMENTS

Rotating elements of the type shown in Fig. 11-3 have been used in radio station KSL in Salt Lake City, Utah.[15] In

Fig. 11-2. (A) The simplest and cheapest way to adjust the reverberation characteristics of a room is to use removable panels. These photographs were taken in the studios of the Far East Broadcasting Company, Hong Kong. (B) Close-up of hanging detail.[42]

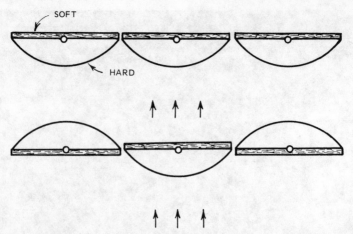

Fig. 11-3. Rotating elements can vary the reverberation characteristics of a room. They have the disadvantage of requiring considerable space to accommodate the rotating elements.

this particular configuration the flat side is relatively absorbent and the cylindrical diffusing element is relatively reflective. A disadvantage of this system is the cost of the space lost which is required for rotation. The edges of the rotating element should fit tightly to avoid coupling between the studio and the space behind the elements.

At the University of Washington a music room was designed with a series of rotating cylinders partially protuding through the ceiling.[15] The cylinder shafts are ganged and rotated with a rack-and-pinion drive in such a way that sectionalized areas of the cylinder exposed give moderate low-frequency absorption increasing in the highs, good low-frequency absorption decreasing in the highs, and high reflection absorbing little energy in lows or highs. Such arrangements, while interesting, are too expensive and mechanically complex to be seriously considered for most studios.

HINGED PANELS

One of the least expensive and most effective methods of adjusting studio acoustics is the hinged panel arrangement of Figs. 11-4 A and B. When closed, all surfaces are hard (plaster, plasterboard, or plywood). When opened, the exposed surfaces are soft. The soft surfaces may be covered with low-density mineral fiber blanket 2 to 4 inches thick. This blanket could be covered with cloth for the sake of appearance. Spacing the mineral fiber from the wall would improve absorption at low frequencies.

VARIABLE RESONANT DEVICES

Resonant structures for use as sound absorbing elements have been used extensively in the Danish Broadcasting House in Copenhagen.[43] One studio used for light music and choirs employs pneumatically operated hinged perforated panels as shown in Fig. 11-5. The effect is basically to shift the resonant peak of absorption as shown in Fig. 11-5B. The approximate dimensions applicable in Fig. 11-5A are: width of panel 2 ft, thickness ⅜ in., holes ⅜ in. diameter spaced 1-⅜ in. on centers. A most important element of the absorber is a porous cloth having the qroper flow resistance covering either the inside or outside surface of the perforated panel.

When the panel is in the open position the mass of the air in the holes and the "springiness" (compliance) of the air in the cavity behind act as a resonant system. The cloth offers a resistance to the vibrating air molecules, thereby absorbing energy. When the panel is closed the cavity virtually disap-

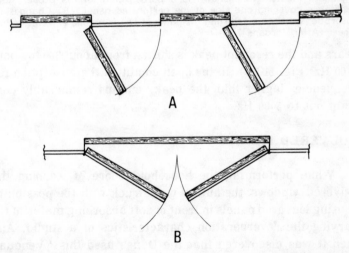

A

B

Fig. 11-4. An inexpensive and effective method of incorporating variability in room acoustics is through the use of hinged panels, hard on one side and absorbent on the other.

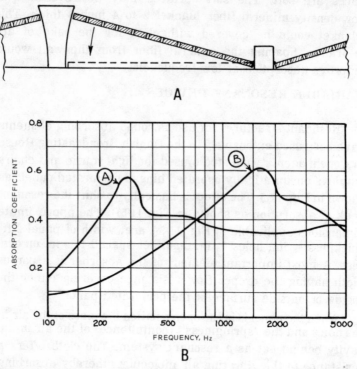

Fig. 11-5. (A) Pneumatically operated hinged, perforated panels used to vary reverberation in the Danish Broadcasting House in Copenhagen. An important element not shown is a porous cloth of the proper flow resistance covering one side of the perforated panel. (B) Changes in absorption realized by shifting the panel of (A) from one extreme to the other. (After Jordan[43])

pears and the resonant peak is shifted from about 300 to about 1700 Hz (Fig. 11-5B). In the open condition the absorption for frequencies higher than the peak remains remarkably constant out to 5000 Hz.

LOUVERED PANELS

While performing the household chore of washing the louvered windows, the author was struck with the possibility of using louvered panels in front of soft absorbing material for varying the reverberation characteristics of a studio. And then it was discovered that the Danes used this "Venetian Blind" method a quarter of a century ago.[11] The louvered panels of an entire section can be rotated by the action of a single lever in the frames commonly available for home construction, Fig. 11-6A. Behind the louvers is a low-density mineral fiber board or batt. The width of the panels determines whether they form a series of slits, Fig. 11-6B, or seal tightly together, Fig. 11-6C. In fact, opening the louvers of Fig. 11-6C slightly would approach the slit arrangement of Fig. 11-6B acoustically, but it might be mechanically difficult to arrange for a precise slit width.

The louvered panel arrangement is basically very flexible. Assuming that the louvers are open, the mineral fiber can be of varying thickness and density and fastened directly to the wall or spaced out different amounts. The louvered panels can be of hard material (glass, hardboard, asbestos board) or of softer material such as wood and they can be solid, perforated, or arranged for slit-resonator operation. In other words, almost any absorption-frequency characteristic we have seen in the graphs of earlier chapters can be matched with the louvered structure with the added feature of adjustability.

THE SNOW ADJUSTABLE ELEMENT

Although snow is a phenomenally good sound absorber, it is neither readily available in many parts of the world nor well adapted to listening room or studio applications. We have in

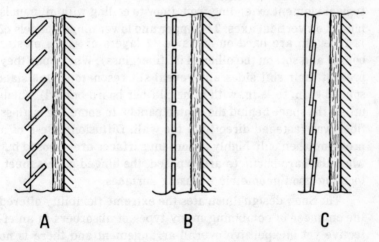

A   B   C

Fig. 11-6. Louvered panels may be opened to reveal absorbent material within, or closed to present a reflective surface. Short louvers can change from a slat resonator (closed) to reveal absorbent material within (open).

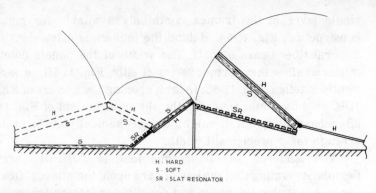

H - HARD
S - SOFT
SR - SLAT RESONATOR

Fig. 11-7. Variable acoustic elements in the mixing-looping stage at Columbia Pictures Corporation, Hollywood. Reflective areas are presented when closed, absorbent areas and slat resonators are presented when the doors are opened. (After Snow [44])

mind here a studio design by the late William B. Snow for the sound mixing-looping stage at Columbia Pictures Corporation studios in Hollywood.[44] Sound mixing requires good listening conditions; looping requires variable voice recording conditions to simulate the many acoustic situations of motion picture scenes portrayed. In short, reverberation time had to be adjustable over about a two-to-one range for this 80,000 cu ft stage.

Both side walls of the stage were almost covered with the variable arrangement of Fig. 11-7 which is a cross section of a typical element extending from floor to ceiling with all panels hinged on vertical axes. The upper and lower hinged panels of 12 ft length are hard on one side (2 layers of ⅜ in. plaster board) and soft on the other (4 in. fiberglass). When open they present their soft sides and reveal slit resonators (1x3 slats spaced ⅜ in. to ¾ in. with mineral fiber board behind) which utilize the space behind the canted panels. In some areas glass fiber was fastened directly to the wall. Diffusion is less of a problem when only highly absorbent surfaces are exposed but when the hard surfaces are exposed, the hinged panels meet, forming good geometric diffusing surfaces.

The Snow design illustrates the extreme flexibility offered the engineer in combining many types of absorbers in an effective yet inexpensive overall arrangement and there is no reason why the hi-fi enthusiast couldn't use them just as effectively with a bit of care.

TUNING
THE LISTENING ROOM

12

The man in the listening booth listens intently to the sound coming over his loudspeaker and many decisions are based upon what he hears. Is he hearing what he should hear? If the same sound is routed to another monitoring room will he hear the same thing? Will another person listening to this material in his living room hear something quite different? Will he be able to understand the spoken words? Will music have a well-rounded fullness?

The operator in the control room uses his program equalizer to make what he hears over his monitor speaker sound more like what he hears live in the studio. Is he merely compensating for deficiencies in his studio and monitoring system or is he making changes really needed to improve the product? The fidelity of the monitoring room reproducing system is vitally important. The hi-fi enthusiast is interested in this same quality.

THE LISTENING PROBLEM

The general practice has been to buy amplifiers on the basis of flatness of response and other features. This flat response can be easily realized under operating conditions. The loudspeaker is also commonly selected on the basis of response curves supplied by the manufacturer. These curves are obtained in anechoic chambers which are quite different from the room in which the loudspeaker will be used. The acoustics of the space and the loudspeaker operating in that space have been the neglected links in the high fidelity chain.[45-49]

THE LISTENING CHAIN

Organ builders have long recognized the importance of adapting a new organ to the space in which it is being in-

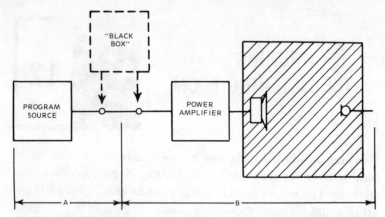

Fig. 12-1. The listening chain may be broken down into the (A) and (B) parts. Section B includes the power amplifier, the loudspeaker and the acoustical environment of the listening space. Section B presents the greater problems.

stalled. Careful adjustment of the sound level output of each pipe is a process they call "voicing" which makes the new organ sound right in that auditorium. The recognition of the interaction of the environment and the sound radiator is the principle we wish to apply to our hi-fi listening or monitoring room.

It is helpful to consider our "listening chain" in two separate but, hopefully, compatible parts as shown in Fig. 12-1. The program source can be live, recorded, or broadcast material. In studio operation, Section A would be considered to end at the program bus or zero level line. In home hi-fi installations Section A would end at the power amplifier input. Section B encompasses the power (monitor) amplifier, the loudspeaker, and the acoustical environment of the listening space. The listener himself is replaced by a calibrated microphone placed where his ears are normally located.

We consider the power amplifier and the loudspeaker together because of the necessity of careful matching of the one to the other. This amplifier should have ample power and an essentially flat characteristic. The object of our quest is to determine what the "black box" should contain to give the overall listening chain a flat response. In the past, problems centering in this acoustical link were either ignored or patiently endured, or solutions were approached on a haphazard and subjective basis. If we can determine what should be in the black box, we will have made a great step forward toward true fidelity of reproduction.

ACOUSTIC RESPONSE

The response of the electronic portion of A in Fig. 12-1 can be readily determined by a variable frequency oscillator or even a standard test disk or tape and a sensitive voltmeter, such as a VU meter. An attempt to determine the response of Section B in Fig. 12-1 by feeding pure tones from an oscillator into the power amplifier would be messy because slight changes in frequency result in major changes in the microphone output due to standing waves in the room. At a fixed pure tone frequency slight changes in the position of the microphone would also result in changes in level which would be difficult to interpret. One solution is to "warble" the tone, or vary the frequency rapidly a small amount above and below the frequency under study (for example, plus and minus 10 percent with a modulation frequency of about 6 Hz). In this way meter readings can be somewhat tamed. However, experience has shown that bands of pink noise (see Chapter 3) are far more satisfactory for such measurements. Bands of one octave, half-octave, or one-third octave width are most commonly used.

Let us assume that the microphone in the listening room of Fig. 12-1 is the microphone of a calibrated sound level meter (SLM). We let this SLM represent our listener because we can get numbers out of it whereas we can get only subjective judgments out of the listener. As we feed a one-third octave band of noise centered on, say, 500 Hz, into the input of the power amplifier (making sure the impedance relationships are not disturbed) we note the reading on the SLM. This procedure is repeated for other bands of noise at higher and lower frequencies, **but always with the input held constant** as shown in Fig. 12-2A. Fig. 12-2B is a plot of the SLM readings for the corresponding center frequencies of the noise bands. This is called the **acoustic response** of the B section of Fig. 12-1 and a considerable deviation from the flat condition is noted. This is sometimes referred to as the "house curve." As the power amplifier is essentially flat, this deviation is chiefly due to the loudspeaker and room and their effect on each other. We are

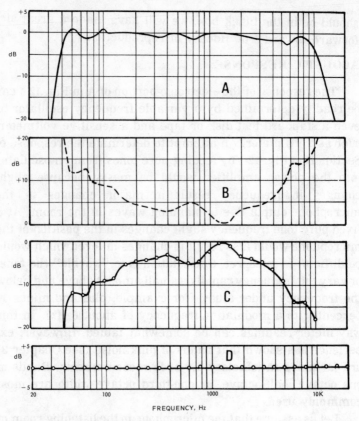

FREQUENCY, Hz

Fig. 12-2. Steps in obtaining the acoustic response of a listening room and equalizing it.[50]
(A) Input to power amplifier is held constant.
(B) Sound level meter readings from microphone of Fig. 12-1.
(C) Inverse of (B) which is the equalization required to correct the variations from flatness.
(D) The equalized listening room at the position of the ear of the listener.

really not much interested in dividing the blame but will proceed to correct for both together.

Fig. 12-2C is the inverse of the acoustic response graph(B) which is required to flatten the fluctuations in B. An equalizer adjustable in one-third, one-half, or octave intervals can be made to approximate C; the narrower the band, the closer the approximation. This is the equalization that must be placed in the black box between Sections A and B in Fig. 12-1 in order that the overall characteristics be made reasonably flat as in Fig. 12-2D.

The graphs of Fig. 12-2 have been obtained from actual data taken in an 11,400 cu ft motion picture rerecording (mixing) studio using one-third octave filters. Actually, a sound analyzer and graphic level recorder were used.[50]

Some experts (notably those correcting motion picture theaters) prefer octave intervals because of the degree of smoothing resulting from their use. Others prefer one-third octaves (fancy hi-fi installations, recording studios, auditoriums) so that almost every squiggle of the house curve can be ironed out. The equalization filters used for the correction are often of the active type (involving solid-state amplifiers and networks) in the budget operations but the more expensive passive filters utilizing inductors and capacitors are usually used in professional installations.[51-54]

LIMITATIONS OF ROOM / SPEAKER EQUALIZATION

Why bother to acoustically treat the listening room? Why not correct for all its acoustic deficiencies by equalization? It sounds nice, but it won't work. Frequency response of the listening chain is the decisive factor in what the listener hears only if: (a) distortion, hum, and hiss are essentially absent in the loudspeaker output, and (b) there are no gross acoustical defects in the listening room. Gross acoustical defects would include serious colorations due to normal modes, inadequate diffusion, improper reverberation time or the presence of flutter. The process of room / speaker equalization may reduce or eliminate colorations due to isolated modes or groups of modes, but the only safe way would be to minimize such colorations before equalization. For example, the 10 dB equalization boost around 100 Hz in Fig. 12-2C might be just the wrong thing for isolated modes near this frequency.

Equalization works wonders in correcting those intangible, hard-to-get-at deficiencies of the room and the speaker-room coupling but cannot cure every acoustical problem of the room or mechanical problem of the loudspeaker.

TONE CONTROLS

Every hi-fi enthusiast has done something in the direction of acoustic response equalization with his treble and bass tone

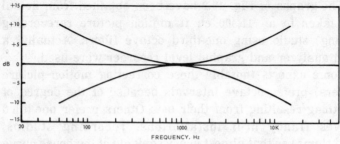

FREQUENCY, Hz

Fig. 12-3. The very limited correction available from the common tone control.

controls. The range of the simplest form of the traditional tone control may be something like that shown in Fig. 12-3. It is obvious that this, even with a midrange control, can do little in correcting a house curve like Fig. 12-2B.

Five-Band Control

A step in the right direction is going from the two- or three-point tone control to one controlling five bands. Typical of this type is the JVC Nivico SEA system (SEA means Sound Effect Amplifier, an unfortunate confusion of terms) which offers plus and minus 10 dB control at 60, 250, 1000, 5000 and 15,000 Hz as shown in Fig. 12-4. A SEA system having seven controlled bands is also on the market. Other companies offering equipment of this type are Harmon Kardon, Metrotec Industries, etc.

Ten-Band Control

Several manufacturers offer 10-band control. These are octave bands with something like plus and minus 12 dB control at the center frequency of each band. The Advent "Frequency Balance Control" and the Soundcraftsmen "Audio Frequency Equalizer" are typical of the 10-band devices which are illustrated in Fig. 12-5.

With octave band control the hi-fi buff has an excellent opportunity to correct for all but the smallest squiggles in his house curve. In fact, Swedish and Danish leaders in the study of the problem of acoustic response of cinema theaters feel that octave bands have distinct advantages over more elaborate systems. Other professionals tend toward the more detailed analysis and equalization.

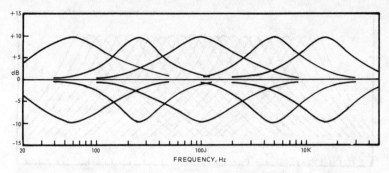

Fig. 12-4. The five-band control offers more flexibility in equalization than the common tone control, but still it is very limited.

Twenty-four-Band Control

In their "Acousta-Voicing" activity, Altec-Lansing takes the fully professional route with sophisticated measuring equipment and one-third octave band-rejection filters. As these are of the passive type, the filters bring all the humps and valleys of the house curve down to a flat condition and the insertion loss is compensated by amplifiers.

The audible band is well dissected by one-third octave filters as shown in Fig. 12-6. These curves can represent the shape of either band-pass or band-rejection filters depending on the arrangement of the dB scale and whether you like them as shown or upside down.

Altec-Lansing carries a heavy program in equalizing auditoriums. One of the important results of this process is the reduction of feedback howling in sound reinforcing systems. In this respect the average auditorium job can be more complex than equalizing a listening room. Altec-Lansing now

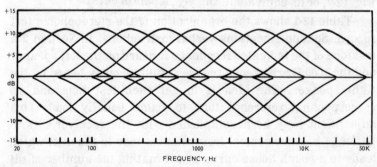

Fig. 12-5. The ten-band control offers a wide range of equalization at octave intervals.

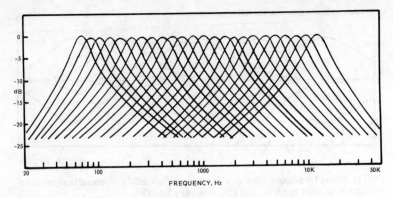

FREQUENCY, Hz

Fig. 12-6. The twenty-four-band control utilizes one-third octave filters and is most flexible for professional acoustic response equalization.

offers a less expensive dual 24-band equalizer, the "Acousta-Voicette," which uses active filters and is intended for the better home hi-fi installations.

A HOME HI-FI EQUALIZING PROCEDURE

Soundcraftsmen*offers a recording with "tones" (bands of pink noise) for testing and full instructions for the hi-fi man to determine his own house curve. The system is basically that of Fig. 12-1 except that the golden ears of the hi-fi man himself may be used instead of an expensive calibrated sound level meter. This has its limitations, but it also tends to correct for any hearing impairment in those golden ears! All that is needed is a good stereo component system, Soundcraftsmen's instructional test record, and a dual-channel octave band equalizer such as the Soundcraftsmen equalizer pictured in Fig. 12-7, or its equivalent. Oh, yes, a pair of ears!

Table 12-1 shows the organization of the stereophonic test record. Section 1 contains verbal instructions. In Section 2 a test tone of the indicated frequency is heard on the left speaker and the 1000 Hz reference tone is simultaneously heard on the right speaker. The idea is to get each test tone and its corresponding reference tone to sound equally loud. This adjustment may be accomplished with the receiver's (amplifier's) balance control from one test tone to the next which leads to a rough house curve by estimating the number of dB

---

* 1320 E. Wakeham Ave., Santa Ana, CA 92705

correction on the balance control to make each test tone equal in loudness to its companion reference tone. If a Soundcraftsmen equalizer (or equivalent) is in the circuit, the appropriate band control is adjusted until the test and reference tones are equal and equalization for that band is accomplished.

After the Section 2 routine has been completed, the left channel is either equalized or the corrections necessary for its equalization are in hand and the similar Section 3 procedure is followed for the right channel. There would be greater accuracy and ease of adjustment if, while adjusting the left channel, the right speaker with its reference tones were alongside the left speaker. For the Section 3 procedure in equalizing the right channel, the left speaker would be moved to a position close to the right speaker. In fact, an even better way is to be able to switch back and forth between test and reference tone on the same speaker. The method of doing this is explained on the Soundcraftsmen record jacket.

If a sound level meter is available, the record can be used as a source of test tones by disconnecting the reference tone channel. Since all the test tones are recorded at a uniform level, the equalizer band controls would be adjusted to give uniform level on the SLM readings for each channel. The "flat" position of the SLM would be used for this purpose to achieve an overall flat system response. This establishes the basic listening condition and further specific equalization adjustments to accomplish specific things can always be added.

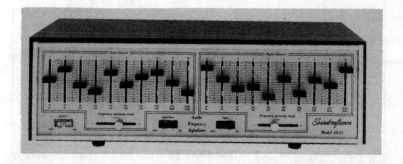

Fig. 12-7. The Soundcraftsmen octave band, two channel equalizer is ideally suited for hi-fi use for the equalization of speaker and listening room. (Soundsraftsmen photo.)

Table 12-1. Contents of Soundcraftsmen Instructional Test Record

Section 1

Verbal instructions

Section 2

Left Channel	Right Channel
Wide band pink noise	Wide band pink noise (20 seconds)
31.5 Hz Test Tone*	1000 Hz Reference Tone**
63 Hz	
125 Hz	
250 Hz	
500 Hz	
1000 Hz	
2000 Hz	
4000 Hz	
8000 Hz	
16,000 Hz	

Section 3

Left Channel	Right Channel
1000 Hz Reference Tone	31.5 Hz Test Tone
	63 Hz
	125 Hz
	250 Hz
	500 Hz
	1000 Hz
	2000 Hz
	4000 Hz
	8000 Hz
	16,000 Hz

*"Test Tone" One octave pink noise centered on indicated frequency, all recorded at same level.
** "Reference Tone" One octave band of pink noise centered on 1000 Hz, level varying according to Fletcher-Munson curves of Fig. 2-2.

EVALUATING STUDIO ACOUSTICS

13

The pioneer scientist was right when he said, "To measure is to know." Only in this way can subjective factors be controlled. However, the very act of hearing is subjective and a trained listener might very well detect flaws in a studio, elegant graphs and sophisticated measurements to the contrary. This does not mean that measurements are worthless. It only means that if one is processing programs for ultimate consumption by the human ear, a trained ear and measurements must supplement each other. The science of acoustics has grown to maturity during the past 100 years but there is still something of an art about its practice.

Things can be much more complicated in acoustical measurements than in, for example, measurements in simple electronic circuits. There are relatively few engineers, physicists, and technicians who have the requisite training, experience, and measuring equipment to design listening rooms and studios, correct acoustic defects, or verify an in-installation.[55] The services of those available in this field come with an understandably high price tag. Is there any hope for the serious hi-fi enthusiast, or those operating small studios beset with budget problems? The answer is "yes" and we shall examine some of the ways it can be done.

HOW EXPERTS EVALUATE STUDIO ACOUSTICS [51-54]

Although the measurement of reverberation time is not everything, it is an excellent beginning and it has valuable fringe benefits. A block diagram of equipment suitable for measuring reverberation time in the professional manner is shown in Fig. 13-1. The loudspeaker, driven by the amplified signal from the signal source, fills the studio with sound as switch S is closed. The amplified signal picked up by the

173

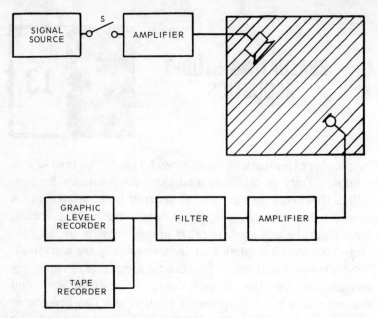

Fig. 13-1. The standard arrangement for measuring the reverberation characteristics of an enclosure.

nondirectional microphone is fed to a graphic level recorder. After the sound in the studio has built up to a steady-state condition, switch S is opened and the graphic level recorder traces the all-revealing decay of the sound in the room. This may be repeated at different frequencies and with the microphone at different locations.

The signal might be a pure tone (with many complicated effects), a "warble" tone, or bands of pink noise.

A good filter in the pickup circuit of Fig. 13-1 will increase the dynamic range of the decay record. A tape recording of the decaying signals at various frequencies and various microphone positions may be kept for later detailed study and as a permanent record of the raw data. In fact, high quality tape recordings are frequently used to minimize the amount of heavy equipment necessary to investigate a room.

The graphic level recorders commonly used in reverberation time measurements take the linear input signal and display it logarithmically on a paper tape, yielding a straight line for the exponential decay as we saw in Fig. 9-2. This record of the decay of sound in the room yields information of great value other than reverberation time. For example, the very smooth decay of Fig. 13-2A indicates good sound diffusion while the irregular decay of Fig. 13-2B indicates poor diffusion. A decay record having a double slope, Fig. 13-2C, usually indicates the existence of a flutter echo between two reflective surfaces.

Unfortunately, graphic level recorders, filters, and other sophisticated instruments are not available to many who are interested in checking out their listening rooms or studios. Lucky ones may have access to such equipment through friends, but most of the readers of this book must rely on their own resourcefulness. Here are some suggestions that may help.

SIGNAL SOURCES

The catalogs of the manufacturers of this type of equipment will probably recommend a random noise generator and a one-third octave filter costing several thousand dollars as the signal source. With one-octave bands of pink noise available on Soundcraftsmen's test record, one can get by quite well without these two instruments.

Another possibility is exciting the room with a noise source covering a wide band (see Fig. 3-4D) and then playing a tape recording of the wideband decay into a graphic level recorder through a set of one-third octave filters. The signal-to-noise ratio may not be quite as good with this method but it has been used successfully. While it simplifies the signal source end, elaborate equipment is still needed to analyze the easily obtained decay records.

Although not recommended for the inexperienced person, impulse noises such as pistol shots (blank cartridges, of

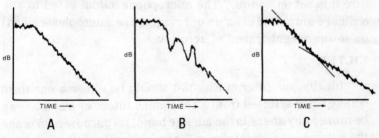

Fig. 13-2. Typical traces of the decay of sound in a room;
(A) normal decay, good diffusion, (B) poor diffusion, and (C) a double slope which usually reveals the presence of a flutter echo.

175

course) can be used. Believe it or not, there is a very respectable paper in an acoustical journal presenting bursting balloons as a source of energy distributed throughout a wide frequency range.[56] The larger balloons available at any "five and dime" have high peak output throughout the audible band, so high in fact that precautions should be taken against ear damage. Great care must be exercised in the interpretation of results with impulsive noise excitation.

AMPLIFIER-LOUDSPEAKER

There should be no problem supplying these items of Fig. 13-1. High acoustical power radiated into the room under test is advantageous as it gives a longer decay trace which yields more information.

A word of warning should be given regarding the possibility of damaging your loudspeaker. The high-frequency driver of the average three-way loudspeaker may be designed to handle only the small amount of power at those frequencies found in ordinary signals. If you hit it hard with a one-third or one-octave band of pink noise within its range, it may go down in flames. Heavy duty two-way speakers are safer and for exciting a room with random noise, speaker quality is unimportant.

MICROPHONE

The reverberant signal picked up by a microphone (Fig. 13-1) arrives from every direction. If the microphone has significant directivity, it will discriminate against some of the incident energy. Reverberation measurements commonly stop at 4000 Hz. Use the smallest microphone available and be sure it is set on "omni." The microphone output is fed to any ordinary amplifier capable of bringing the microphone signal up to the neighborhood of zero level.

FILTER

Ideally, the filter of Fig. 13-1 should be either a one-third octave filter selected from a set of such filters or one that can be moved anywhere in the audible band. Its purpose is to make the equipment which follows it sensitive only to the band fed into the room. In this way the noise outside this band is rejected and the signal-to-noise ratio is improved.

Consider Fig. 13-3, illustrating the rather self-evident fact that there are two ways of improving the signal-to-noise ratio, increasing the signal level, or decreasing the noise level. In Fig. 13-3A the initial signal level is low and the noise level high, a poor combination yielding a short decay of limited value. Within 10 dB of the noise the slope is definitely affected as the noise energy adds to the energy of the decay signal. If the signal put into the room is increased, a greater portion of the decay is rescued from the noise, Fig. 3-13B. If the noise level can be reduced (by doing something like turning off the air conditioning equipment) and the signal level also increased, we have greatly increased the proportion of the decay visible as seen in Fig. 13-3C. Without filters to make a substantial reduction in noise level, it will be necessary to take every possible step to stretch the signal level up and push the noise level down to get usable decay traces.

Fig. 13-4A shows tracings from actual reverberation measurements made by playing a tape of the decays into a graphic level recorder. No filters were available, the signal level into the room was limited, and the noise level was high. In other words, everything was wrong. The 125 and 250 Hz traces are completely unusable. The great fluctuations inherent in random noise of this bandwidth and for these frequencies make a bad situation even more hopeless. But a reasonably good decay slope could have been obtained if we had, say, a 30 dB decay in the clear instead of the 8-12 dB at 125 and 250 Hz. The fluctuations subside as the one-third octave bandwidth (in hertz) increases and above 500 Hz the decays are reasonably linear and measurable. The T_{60} is found by a

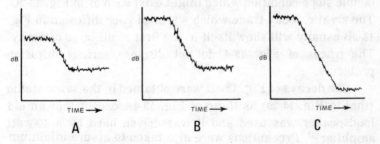

Fig. 13-3. The effect on decay traces of the degree of excitation of the room and the noise level; (A) low initial sound level, high noise level, (B) high initial sound level, high noise level, and (C) high initial sound level, low noise level.

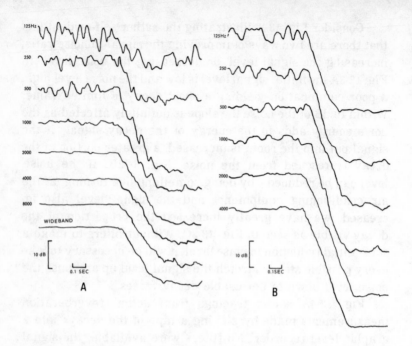

Fig. 13-4. (A) Actual traces of the decay of one-third octaves of noise in the studio of Figs. 14-24 to 14-28 under the conditions of low room excitation and high noise level. (B) The same under conditions of strong room excitation and low noise level. Writing speed 1000 mm/sec in (A), 300 mm/sec in (B).

bit of "similar triangle" geometry. We fit a straight line to the decay. If it takes 0.4 second for the slope of the line to fall 30 dB, the T_{60} is, of course, 0.8 second.

With the shorter decays resulting from working without filters, low room levels and high noise levels, certain information is sacrificed. An example of this is obscruing any double slope condition which might exist such as in Fig. 13-2C. The erratic decay trace which warns of poor diffusion in Fig. 13-2B usually will show itself in the first 30 dB or so of decay. The traces of Fig. 13-4 do not disclose serious diffusion problems.

The decays of Fig. 13-4B were obtained in the same studio (that in Fig. 14-29) as those of Fig. 13-4A except that an old loudspeaker was used and it was driven hard by a 50-watt amplifier[50]. Precautions were also taken to assure minimum noise level. These decays of Fig. 13-4B prove that for small rooms, at least, satisfactory decay traces can be obtained without a filter in the microphone circuit. We can see that diffusion is excellent and there seems to be no evidence of double slope.

GRAPHIC LEVEL RECORDER

Amplifiers, speakers, and microphones can easily be found, even though not of precise, professional pedigree. The signal source of Fig. 13-1 can be octave bands of noise (it doesn't have to be pink for this use, white is just fine) available from a store. Filters we will miss, but we can do quite well without them. But how about the graphic level recorder? How can we stretch out a few tenths of a second and study the decay of sound level over that time? Good graphic level recorders are expensive and no cheap substitutes are known to the author. But there is a way! If a cathode ray oscilloscope is available we can build the poor man's substitute for a graphic level recorder. We need two special attachments, one essential and the other highly desirable. The essential one is a logarithmic amplifier, a number of which are on the market. While the electronic wizard can concoct one of his own, this isn't the place to go into details on the subject aside from a few references.[57,58] The desirable features are linearity between volts in and dB out and a good dynamic range, 50 or 60 dB hopefully, although here again, much can be done with less.

The less essential but highly desirable feature is a triggered-sweep feature on the scope. The switch that is opened to stop the radiation of sound from the loudspeaker can also be arranged to initiate a relatively slow horizontal sweep of the spot across the screen. The amplified reverberant sound being picked up by the microphone is applied to the vertical plates of the scope through the log amplifier. In this way the decay is nicely displayed every time the sound source switch is opened. A Polaroid or other camera with high speed film can give a permanent record of selected decays. Such a setup as used at Moody Institute of Science is shown in Fig. 13-5. In this case the Tektronix Type 561 scope used has a single-sweep, self-triggering mode built in. Fig. 13-6 shows typical decay records obtained.[26] By calibrating the vertical dB scale and the horizontal time scale, values of T_{60} can be at least estimated to verify calculations.

A commercial decay rate meter is available which operates with a scope in a manner similar to that described

Fig. 13-5. Measuring the decay of sound in an enclosure by means of a cathode ray oscilloscope, logarithmic amplifier (vertical plates) and a single sweep arrangement (horizontal plates). Permanent records are made photographically.

(Moody Institute of Science photo, Reprinted with permission from Journal of the Audio Engineering Society.[26])

above but with an interesting method of reading the T_{60} directly from a dial.[59] A capacitor discharge is also an exponential decay which in this instrument is made to appear on the screen as a straight line whose slope is adjustable by varying the RC circuit. By adjusting the RC decay line to coincide with the slope of the room decay, a direct readout of the T_{60} may be obtained over the tremendous range of 0.005 to 10 seconds.

Another instrument on the market measures the time for the reverberant sound level to fall 15 dB, the T_{60} being obtained by multiplying this by four.[60] One-third octave filters are incorporated and it is interesting to note that the one centered on 250 Hz is the lowest band, possibly because of the ever present noise problems at the 125 Hz and other low frequency bands.

REVERB TIME: STOPWATCH METHOD

In the early 1920s Professor W. A. Sabine systematically investigated the reverberation characteristics of the Fogg Art Museum of Harvard University. In this historical first, an

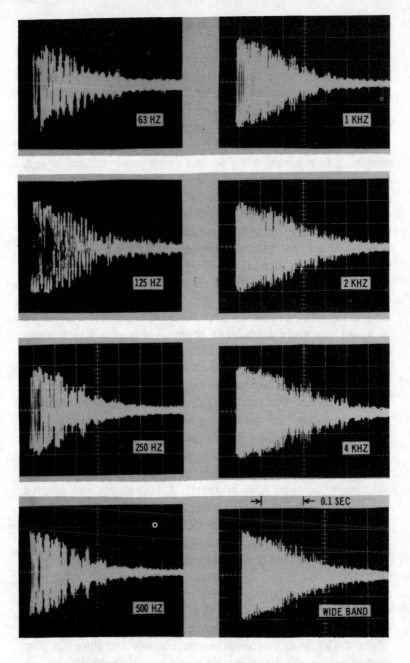

Fig. 13-6. Decay of one-third octaves of noise as recorded with the system of Fig. 13-5. By calibration of vertical and horizontal scales, values of reverberation time can be determined. (Moody Institute of Science photo. Reprinted with permission from Journal of the Audio Engineering Society.[26])

organ pipe (512 Hz) was used as the source of sound. A stop-watch was started the instant the source ceased emitting and was stopped the instant the sound ceased to be audible. As the initial sound pressure level in the room was about a thousand times that of the threshold of hearing, the definition of reverberation time came to be that time for the sound to decay 60 dB. It is amazing how consistent Sabine's measurements were over a period of time and from observer to observer. It is one thing to measure a 5.5 second reverberation time (Fogg Art Museum, empty) by this aural-stopwatch method and quite another to measure a T_{60} of 0.4 second (speech recording studio or living room). The reverberation time of larger studios, however, can be estimated by this method. Dr. William E. ("Ted") Haney has related to the writer how he and his associates used the method to check the studio of Fig. 14-6 having a volume of 27,500 cu ft. The room was filled with sound from a loudspeaker driven by an oscillator and suitable amplifiers. Measurements were made at frequencies selected to avoid the more prominent room resonances which were quite apparent as the oscillator was swept over the band.

A scope was used as a visual indicator rather than depending on the ear as Sabine did. The loudspeaker level was adjusted so that the amplified microphone signal read some arbitrary value such as 3 volts at the scope terminals. The scope gain control was then adjusted so that a signal 60 dB lower (0.003 volt) would give a certain known deflection. With this setting, of course, the trace was far off screen with the 3 volt input, but this is immaterial. A stopwatch was started as the switch was opened and stopped as the trace crossed the 0.003 volt line. (The human reaction time effect tends to cancel out because the reaction time in punching the stopwatch at the start is more or less the same as that at the end of the decay.) With three observers, each making 10 measurements for each frequency, Dr. Haney estimated 10 percent accuracy for reverberation times as short as 0.8 second. This method is not accurate for shorter T_{60}s. The reverberation characteristics of the CAVE studio determined in this way are shown in Fig. 13-7 in which the "before" (A) and the "after" (B) graph are shown. These measurements were started after ¼" fiber board panels were mounted on one wall, hence the lower reverberation time in the low frequencies in A. It was discovered that directivity patterns of the microphones used had considerable effect on the T_{60} values. Nondirectional microphones are required to catch sound arriving in random directions equally well.

EVALUATION OF BACKGROUND NOISE

While discussing the penetration of extraneous noises through thin-walled recording studios in the Orient, a missionary broadcaster said, "...but outside sounds of dogs barking and voices of children at play only add character and interest to a program!" While related sound effects add a tremendous dimension of reality, sanity and rationality demand that we follow the plan of minimizing unrelated interfering sounds.

The ear is well adapted to evaluating background noise; in fact, it is the final arbiter. It is a common experience to hear interfering noises in a studio which cannot be heard on recordings made from that studio because the outside noise is masked by the ever-present system noise. There is little point to working hard to eliminate noise which the ear cannot hear in the final recording.

Illumination fixtures are often a source of buzz or hum, often at twice the frequency of the power source. The filaments of incandescent lamps can generate audible as well as radio frequency interference. Fluorescent fixtures are a

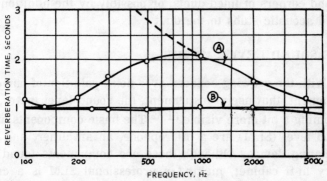

Fig. 13-7. Reverberation time measurements made in CAVE studio in Brazil by stopwatch method. A cathode ray oscilloscope trace was used to indicate when the sound had dropped 60 dB; (A) before treatment, (B) after treatment.

very common source of troublesome buzz, usually traceable to loose laminations in the ballast reactors. It is strongly urged that fluorescent ballast reactors be banished to a box well removed from the recording area.

Air conditioning equipment is notorious as a producer of noise. Chapter 7 points out that such sounds reach the studio either by an airborne or a structure-borne path, or possibly, a complex combination of both. Let us assume that a ventilating fan noise is giving trouble in a studio. How does one determine whether it comes via air or structure? If the whole studio shakes to the tune of the fan motor, it may be obvious that it is a building vibration problem requiring a resilient mounting for the machinery. On the other hand, the transmission path by which the sound reaches the studio might be more difficult to determine.

One way of solving such a problem is to record the sound close by the offending source and play this back on a loudspeaker placed in the equipment room but with the machinery silent. The level is adjusted to approximate that which the machinery would produce if it were running. Now we have a producer of the noise which is not fastened to the structure. If the noise radiated by the loudspeaker is **not** heard in the studio, the problem should be attacked on a structure-borne basis. One should also be alert to the possibility of fan blade modulation noise being transmitted through the air ducts. This can be reduced by lining the ducts with mineral fiber board, by routing the air around absorbent baffles in a noise trap, around corners of lined ducts, or possibly by the addition of tuned acoustic stubs to the duct.

THE SOUND LEVEL METER

With the growing concern over noise pollution of our environment, that useful instrument, the sound level meter, is becoming a bit more visible.[61,62] The basic components of a sound level (SLM) are a microphone, an amplifier, and an indicating device. Although these are components found in every hi-fi cabinet, making a professional SLM is a considerably more involved undertaking. The microphone must have a flat response, excellent stability and linearity, low distortion, good conversion efficiency, rugged construction, and must come with a certified calibration. The amplifier must incorporate various filters and weighting networks which shape the overall response to simulate that of the human ear at several loudness levels. The readout meter must be designed to present root-mean-square (rms) readings accurate for any wave shape of sound being measured. The better the SLM, the lower the sound levels it can read.

The weighting networks of precision SLMs are an attempt to make the meter readings roughly comparable to the loudness value of the sounds. The equal loudness contours of the ear dictate the response shape of the weighting networks. The A scale response corresponds to the ear's sensitivity at 40 phons, the B scale at 70 phons, and the C scale is close to flat, a response approached by the ear at the higher loudness levels. The response of the A, B, and C scales is shown in Fig. 13-8 and they are derived from the equal loudness contours of Fig. 2-2.

Sound level meters can perform many useful services around a studio. Interfering noises can be measured, removing uncertainties of judgment based on listening and variations from one listener to another. It is excellent for measuring the improvement in background noise level as remedial steps are taken. It should be emphasized, however, that only the best (most expensive) SLMS are capable of measuring these small studio noise levels. Measuring the

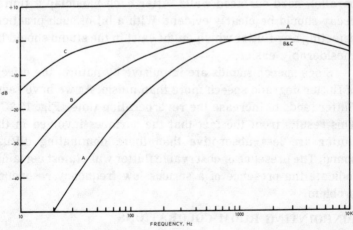

Fig. 13-8. The three response curves of the standard sound level meters. The A curve corresponds to the ear's response at a level of 40 phons, B at 70 phons, and C at high levels.

acoustic response of listening rooms, as we learned in Chapter 12, is today a very important use of SLMs.

If $300 to $700 is not available for purchase of a professional SLM, must acoustic response measurements of listening and control rooms be abandoned? As the high sensitivity, stability, and accuracy of the expensive meters are not essential in the more casual acoustic response checks, the $39.95 Realistic "Music/Sound Level Meter" offered by Radio Shack can serve quite well. It has a range from 60 to 116 dB (see Table 1-2) in 5 ranges, with an accuracy of plus or minus 2 dB and a passband flat from 40 to 14,000 Hz within plus and minus 2 dB. With this meter and the Soundcraftsmen Instructional Test Record, octave band acoustic response measurements are available to the hi-fi enthusiast. Comparing the pink noise band levels into the power amplifier with those read in the listening room on the SLM, you can obtain a house curve such as Fig. 12-2B.

FLUTTER ECHOES

Highly reflective parallel walls tend to give rise to flutter echoes which often may be detected by the unaided ear. A clap of the hands provides an excellent impulsive sound for exciting a flutter mode. Before trying this in a relatively dead studio, it is well to become familiar with the procedure where the flutter is very prominent, such as a large room with parallel, hard surfaced walls. After each handclap a flutter decay should be clearly evident. With a bit of such practice, detecting any flutter which might exist in the studio should be considerably easier.

Since speech sounds are impulsive by nature, the effects of flutter degrade speech more than music. As we have seen, flutter tends to increase the reverberation time (Fig. 13-2C). This results from the fact that the surfaces involved in the flutter are less absorptive than those dominating diffuse sound. The presence of observable flutter will almost certainly indicate the presence of a serious low frequency resonance problem.

PINPOINTING ROOM COLORATIONS

Applying a good critical ear to male speech picked up from a room over a high quality system is one quite effective

organ pipe (512 Hz) was used as the source of sound. A stop-
way to detect axial mode colorations. Only the more
prominent colorations can be singled out by this method,
however, and it is difficult to estimate the precise frequency at
which they occur so that corrective measures may be taken.

The BBC has used an interesting device to assist the ear, a
tunable amplifier that amplifies a narrow frequency band (10
Hz) about 25 dB above the rest of the spectrum. A small
proportion of the output of this selective amplifier is mixed
with the original signal so that its contribution is scarcely
noticeable except when tuned to the frequency of a coloration.
In this way the colorations are easily detected and their
frequency determined.

Using this instrument in many treated studios disclosed
the fact that one or two colorations in a studio were quite
common and that most of them were between 100 and 175 Hz
with a smaller number around 250 Hz as we saw in Fig. 5-4.

EVALUATING SPEECH INTELLIGIBILITY

The so-called "articulation" test is an example of quan-
titative measurements without sophisticated and expensive
equipment. It is a measure of the intelligibility of speech
originally developed for testing transmission over telephone
equipment but it can be applied equally well to mouth to ear
communication in a room or over any sound reproducing
system. Fundamentally, speech sounds are pronounced into
one end of a transmission system and an observer writes down
the sounds heard at the receiving end. **Percentage articulation**
is the percentage of speech sounds heard correctly when
applied scientifically by a number of speakers and listeners
and using standard word lists. The percentages are in-
terpreted as follows:

Articulation   Understandability

96 percent   "Perfect" hearing, the meaning of the few lost
               words are clear from the context.
85-96 percent Highly satisfactory
75-85 percent Satisfactory
65-75 percent Speech understandability with strained at-
               tention.
Below 65 percent   Unsatisfactory

Words of one syllable are most difficult to understand. Standard word lists and test procedures are available for those interested in following up on such tests.[63] Articulation is affected by the loudness level, noise level, reverberation, and test conditions.

A PICTORIAL TOUR OF STUDIOS AROUND THE WORLD

This tour is conducted for only one reason, to catch a glimpse of features and ideas which might be of interest both to the hi-fi man and the studio operator. We are going to visit some midgets and some giants but in none of them do we indulge in anything like a comprehensive coverage. The amateur can learn much from the professional and, who knows? perhaps the professional can learn something from the little fellow way out in Timbuctu struggling along with little sympathy and even less money.

There is a prodigious amount of recording going on in the "Third World," generally for Christian radio use, by a corps of people whose special training for the job often falls far short of their enthusiasm and dedication. They share with the average hi-fi enthusiast that gnawing compulsion to improve the quality of their work even though there is rarely money enough to go first class.

So, keep your eyes open. Maybe there are a few ideas in the following photos and captions which will be worth the price of admission!

Figs. 14-1 and 14-2. Doors with sufficient transmission loss may be a problem if money is scarce. The Far East Broadcasting Company in Manila is very happy with the inexpensive doors they have developed over a period of years, according to Mr. Hiley Rainer, Projects Engineer. In these general views the magnetic seals of the type commonly used on home refrigerator doors are seen as while lines. For the sides and the top, the seals are mounted to the door casing with the companion iron straps on the door. At the borrom the sealing strip is mounted on the door with the iron strap on the raised threshold to minimize foot damage.

Figs. 14-3 and 14-4. Heavy freezer door hinges are utilized to swing the heavy doors. The magnetic sealing strips and iron straps (refer to Chapter 7) are clearly seen at top, side, and bottom edges of the door and frame.

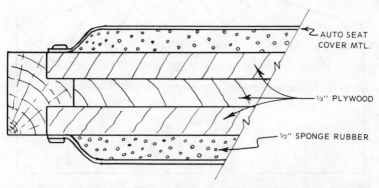

Fig. 14-5. The door is made of three sheets of ½ inch plywood glued together and fitted with a mortised edging. Both sides are covered with ½ inch sponge rubber which, in turn, is covered with automobile seat cover material. (Far East Broadcasting Company photos)

Fig. 14-6. The studio of CAVE (pronounced "Cah'vay") of Centro-Audio-Visual-Evangelico in Campinas, S.P., Brazil was designed by Dr. William E. ("Ted") Haney. First we note that the polycylindrical diffusers are of assorted sizes, the larger ones being more effective as diffusers at the lower frequencies. The drapes absorb high frequency energy when closed and reveal highly reflective wall surface or corrugated asbestos roofing material when retracted, giving some control of the treble reverberation time. (Moody Institute of Science photo.)

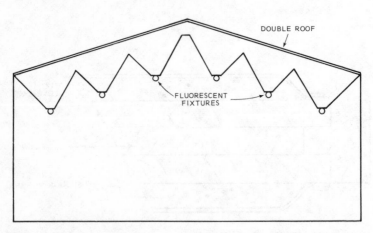

Fig. 14-7. The serrated ceiling in the CAVE studio effectively prevents flutter between it and the floor. The absorption of the ceiling is high because much of the sound undergoes several reflections between adjacent surfaces of a given serration.

Fig. 14-8. The AVEC (Audio-Visual-Evangelism-Committee) studio of the Hong Kong Christian Council in Hong Kong. This studio was designed by Mr. Delbert Rice of Silliman University, Philippines. Polycylindrical diffusers much shorter than wall height dominate the room. These provide needed low frequency absorption and their large chord dimensions would suggest that they are effective diffusers to quite low frequencies. Some of these polys hide building support columns.

Fig. 14-9. The concave ceiling, designed to discourage floor-ceiling reflections, is probably not concave enough to cause focusing problems.

Fig. 14-10. This louvered structure which dominates one wall is reported to be built for low frequency absorption. It is not clear how this structure is supposed to accomplish this. It certainly could be expected to have an absorption and diffusing effect as described in Fig. 14-11.

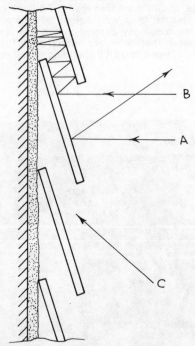

Fig. 14-11. The louvered structure of Fig. 14-10 provides a combination of diffusing and an acoustical "sink" or trap. Horizontal ray A is reflected from the wooden louver externally while ray B is subjected to multiple reflections within the trap. Reflections C from the floor would tend to be completely absorbed.

Fig. 14-12. A studio used primarily for language dubbing operated by Cine Matografica InterAmerica, S.A., Mexico City. The studios were designed by Michael Rettinger, well known acoustics consultant. The slat resonators have slits of different widths, presumably to broaden the peak of absorption. (Moody Institute of Science photo.)

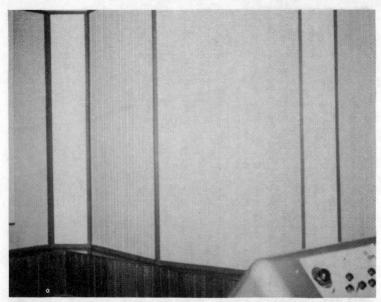

Fig. 14-13. The walls are of canted panels. Some panels are slat resonators, some of perforated hardboard which covers mineral fiber, and some are highly reflective surfaces. Both the angular surfaces and the distribution of different panels contribute to sound diffusion in the room.

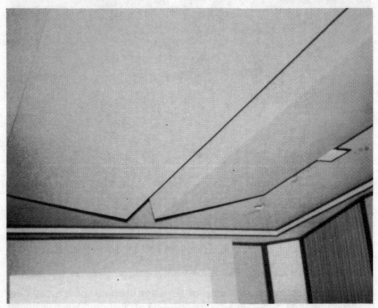

Fig. 14-14. A simple plywood-faced geometrical ceiling feature supplies some low frequency absorption as well as serving to break up multiple floor-ceiling reflections. (Moody Institute of Science photos.)

Fig. 14-15. A voice recording studio at Moody Institute of Science, Whittier, California. The slat resonators on the walls are designed for peak absorption at several low frequencies by varying the depth of the boxes. See Fig. 14-18 for a rear view of one of these absorbers.

Fig. 14-16. The pyramidal ceiling feature contributes by (1) breaking up reflections between floor and ceiling, (2) reducing the effective area of already installed acoustic tile, (3) adding to low frequency absorption by the panel effect, and (4) covering air inlet duct, the absorbent materials laid on its upper surface tending to absorb airborne fan noise coming through the duct. (Moody Institute of Science photos.)

Fig. 14-17. Canted panels are joined to slat absorbers. The wall opposite this one employs the same elements but reversed so that slats face canted panels.

Fig. 14-18. Rear view of low frequency slat absorber such as those mounted on the wall in Fig. 14-15. The mineral fiber board is semirigid, requiring little mechanical support.

Fig. 14-19. The method of constructing the slat absorbers below the polys of Fig. 6-7. (Moody Institute of Science photos.)

Fig. 14-20. The studio of the American Assemblies of God, located in Sha Tin, New Territories, Hong Kong, is used for recording radio programs. The dimensions are approximately 12'x 17'x 7'4'' with the control room window set diagonally across one corner. This very small studio was designed by Mr. L. John Wheatley of the Far East Broadcasting Association of England. Mr. Wheatley has utilized the modular approach, basically, but with 2'' x 8'' lumber on edge forming the modules completely hidden by the covering materials on the ceiling and all walls. On this west wall a drapery can be drawn over the modules for limited control of reverberation.

Fig. 14-21. East wall of the same studio showing the angled control room wall. The air conditioning duct is upper left. Three types of modules are employed with perforation percentages of 5.5 percent, 0.12 percent, and 0 percent. (Hong Kong Baptist College photos.)

Fig 14-22. The door is in the south wall. At a doorway, the built-out wall is apparent.

Fig. 14-23. An effort has been made to have modular sections of one type opposing sections of another type on the opposite wall. Estimations indicate that the reverberation time is close to the optimum value of 0.3 second for a volume of 1560 cu ft. (Hong Kong Baptist College photos.)

Fig. 14-24. A large poly dominates one end of the small Hong Kong Baptist College studio which is used for instruction in radio production. This poly contributes to the low frequency absorption as well as to the diffusion of sound in the studio.

Fig. 14-25. A view of the poly before the outer "skin" was applied. The frame and bulkheads were sealed to the plaster wall with a caulking compound, separating each compartment from its neighbors and from the studio space. Mineral fiber batts are mounted in each compartment. The skin is ¼" plywood to which ⅛" finish veneer has been glued.

Fig. 14-26. Three removable panels covered with Johns Manville ¾" Tempertone tile on one side and ¾" plywood on the other provide some control of reverberation. (Hong Kong Baptist College photos.)

Fig. 14-27. A suspended ceiling 19" below the plaster ceiling provides good absorption throughout the band. This allows the use of practical vinyl tile on the floor without flutter problems. The reverberation decay traces of Fig. 13-4B were recorded in this studio and the actual measured values of reverberation time are:

Frequency	T_{60} (seconds)
125	0.35
250	0.37
500	0.38
1000	0.35
2000	0.47
4000	0.44

These measurements have revealed that a small amount of additional absorption is needed in the 2000 to 4000 Hz region to flatten the T_{60} graph.

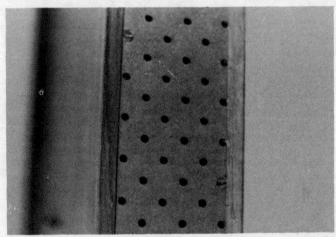

Fig. 14-28. Absorbent door edge to help in trapping outside sound passing through the inevitable crack between door and jamb. This door is constructed according to the sketch of Figs. 7-10 and 7-11, including the sand loading. (Hong Kong Baptist College photos.)

Fig. 14-29. The double glass windows between control room and studios are patterned after the design of Fig. 7-9. (Hong Kong Baptist College photos)

Figs. 14-30, 14-31 and 14-32. Views of the music studio at Radio Hong Kong. Modules after the general BBC plan are used in the acoustic treatment of this studio. Note the various depths of modules employed. It is understood that modules of three different absorption patterns are used.

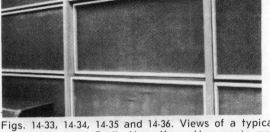

Figs. 14-33, 14-34, 14-35 and 14-36. Views of a typical smaller general purpose studio at Radio Hong Kong. Here again modules are used effectively in achieving proper acoustic conditions. Radio Hong Kong studios and treatment were designed by Mr. Ian Campbell, Lecturer in Architecture, University of Hong Kong. (Hong Kong Baptist College photos.)

Fig. 14-37. Studio 1 at Bayerischer Rundfunk in Munich, West Germany. The very functional walls of this large studio are designed around acoustically absorbent modules. The distribution of these modules in a more or less random configuration aids diffusion of sound. (Photos by FOTO-SESSNER courtesy Bayerischer Rundfunk.)

Fig. 14-38. The acoustic conditions of this room are dominated by shaped and spaced Hunter-Douglas aluminum panels backed by an acoustic pad. This arrangement yields absorption in the 125-1000 Hz region. (Photos by FOTO-SESSNER courtesy Bayerischer Rundfunk.)

Fig. 14-39. A view of another control room at Bayerischer Rundfunk. The studio in the distance is also treated acoustically by the module method. Notice the perforated panel type of absorber to control resonances between the glass plates of the window.

APPENDIX I
SELECTED ABSORPTION
COEFFICIENTS

SELECTED ABSORPTION COEFFICIENTS [17]							
Material	125 Hz	250 Hz	500 Hz	1000 Hz	2000 Hz	4000 Hz	Ref. (Text Figs.)
POROUS TYPE							
Drapes: cotton 14.7 oz / sq yd							
Draped to 7 / 8 area	0.03	0.12	0.15	0.27	0.37	0.42	17 (Fig. 8-2A)
Draped to 3 / 4 area	0.04	0.23	0.40	0.57	0.53	0.40	17 (Fig. 8-2B)
Draped to 1 / 2 area	0.07	0.37	0.49	0.81	0.65	0.54	17 (Fig. 8-2C)
Drapes: Medium velour 14 oz / sq yd draped to half area	0.07	0.31	0.49	0.75	0.70	0.60	31 (Fig. 8-1B)
Drapes: Heavy velour 18 oz / sq yd draped to half area	0.14	0.35	0.55	0.72	0.70	0.65	31
Carpet: heavy, laid on concrete	0.02	0.06	0.14	0.37	0.60	0.65	31 (Fig. 8-1C)
Carpet: heavy, laid on 40 oz hairfelt	0.08	0.24	0.57	0.69	0.71	0.73	31
Carpet: laid on foam or sponge rubber underlay	___	0.05	0.20	0.40	0.60	0.65	32
Carpet: indoor-outdoor	0.01	0.05	0.10	0.20	0.45	0.65	40
Cellulose tile: Simpson Pryotect Tile, standard drilled, ½" thick, cemented to wall.	0.05	0.20	0.56	0.95	0.93	0.74	31 (Fig. 8-1A)
Cellulose tile: Johns-Manville Spintone, ½" cemented to wall	0.09	0.23	0.62	0.75	0.77	0.77	31 (1968)

Material	125 Hz	250 Hz	500 Hz	1000 Hz	2000 Hz	4000 Hz	Ref. (Text Figs.)
MISC BUILDING MATERIALS							
Concrete block: coarse	0.36	0.44	0.31	0.29	0.39	0.25	31 (Fig. 8-1D)
Concrete bock: painted	0.10	0.05	0.06	0.07	0.09	0.08	31 (Fig. 8-1D)
Floor: concrete	0.01	0.01	0.015	0.02	0.02	0.02	31
Floor: linoleum, asphalt rubber or cork tile on concrete	0.02	0.03	0.03	0.03	0.03	0.02	31
Floor: wood	0.15	0.11	0.10	0.07	0.06	0.07	31
Glass: large panes of heavy glass	0.18	0.06	0.04	0.03	0.02	0.02	31
Glass: ordinary window	0.35	0.25	0.18	0.12	0.07	0.04	31
Plaster: gypsum or lime, smooth finish on tile or brick	0.013	0.015	0.02	0.03	0.04	0.05	31
Plaster: gypsum or lime, smooth finish on lath	(a) 0.14 *	0.10	0.06	0.05	0.04	0.03	31
	(b) 0.02 *	0.02	0.03	0.04	0.04	0.03	31 (1968)

* The coefficients listed under (a) are the ones listed in the 1971-72 edition of reference 31, those under (b) were in the previous edition. The examples of Chapter 9 were computed with values from (b) although those in (a) should be used henceforth.

Material	125 Hz	250 Hz	500 Hz	1000 Hz	2000 Hz	4000 Hz	Ref. (Text Figs.)
RESONANT ABSORBERS							
Plywood paneling: 3/8" thick	0.28	0.22	0.17	0.09	0.10	0.11	31 (Fig. 8-11D)
Polycylindrical: Chord 45", ht 16", empty	0.41	0.4	0.33	0.25	0.2	0.22	17 Fig. 8-5A)
Chord 35", ht 12", empty	0.37	0.35	0.32	0.28	0.22	0.22	17 (Fig. 8-5B)
Chord 28", ht 10", empty	0.32	0.35	0.3	0.25	0.2	0.23	17 (Fig. 8-5C)
ditto, filled	0.35	0.5	0.38	0.3	0.22	0.18	17 Fig. 8-5C)
Chord 20", ht 8", empty	0.25	0.3	0.33	0.22	0.2	0.21	17 (Fig. 8-5D)
ditto, filled	0.3	0.42	0.35	0.23	0.19	0.2	17 (Fig. 8-5D)
Perforated panel: 5/32" thick, 4" depth air space, 2" mineral wool							
Percent perforation:							
0.18	0.4	0.7	0.3	0.12	0.1	0.05	17 (Fig. 8-7A)
0.79	0.4	0.84	0.4	0.16	0.14	0.12	17 (Fig. 8-7B)
1.4	0.25	0.96	0.66	0.26	0.16	0.1	17 (Fig. 8-7C)
8.7	0.27	0.84	0.96	0.36	0.32	0.26	17 (Fig. 8-7D)
8" depth, 4" mineral wool Percent perforation:							
0.18	0.8	0.58	0.27	0.14	0.12	0.1	17 (Fig. 8-8A)
0.79	0.98	0.88	0.52	0.21	0.16	0.14	17 (Fig. 8-8B)
1.4	0.78	0.98	0.68	0.27	0.16	0.12	17 (Fig. 8-8C)
8.7	0.78	0.98	0.95	0.53	0.32	0.27	17 (Fig. 8-8D)

Material	125 Hz	250 Hz	500 Hz	1000 Hz	2000 Hz	4000 Hz	Ref. (Text Figs.)
Perforated panel absorber							
with 7" air space plus 1" mineral fiber of 9-10 lb per cu ft density, ¼" cover:							33
Wideband-25 percent perf or more	0.67	1.09	0.98	0.93	0.98	0.96	(Fig. ★ 8-11A)
Mid-peak - 5 percent perf	0.60	0.98	0.82	0.90	0.49	0.30	33 (Fig. ★ 8-11B)
Lo-peak - 0.5 percent perf	0.74	0.53	0.40	0.30	0.14	0.16	33 (Fig. 8-11C)
Perforated panel absorber with 2" space filled with 9-10 lb per cu ft density mineral fiber:							
0.5 percent perf.	0.48	0.78	0.60	0.38	0.32	0.16	33 (Fig. 10-8)
Perforated Transite, Johns-Manville, 3/16" thick, 550 3/16" holes per sq ft on 1" battens with 1" Microlite pad of 0.6 of 0.6 lb density.	0.09	0.31	0.56	0.93	0.68	0.23	31
Drop ceiling: Johns-Manville Acousti-Shell TF, 1/8" thick, 16" airspace.	0.70	0.69	0.66	0.80	0.84	0.83	31 (1968)

    ★ The graphs of Figs. 8-11A & 8-11B are smoothed somewhat from these data.

APPENDIX II

EYRING'S FORMULA
FOR DEAD ROOMS

The Sabine reverberation formula, used for calculating the examples of Chapter 9, is accurate for diffuse conditions and for live rooms. Carl F. Erying derived a formula which is more accurate for relatively dead rooms such as the listening rooms and studios in which we are interested. The Eyring formula can be written to look very much like the Sabine formula but containing a very significant difference:

$$T_{60} = \frac{(0.05)\ (V)}{(S_T)\ (A)}$$

where

T_{60} = reverberation time, seconds
V = volume of room, cu ft
S_T = total surface area of room, sq ft
A = factor derived from a_{ave} (average absorption coefficient) of Equation 9-1 with the help of the graph of Figure A in this Appendix.

Computing T_{60} from the Eyring formula is a bit more tedious than the Sabine formula, but its improved accuracy for small, dead rooms makes it useful for serious calculations. Following through on a specific example will illustrate the necessary steps.

Let us apply the Eyring formula to the hi-fi living room, Example 4 in Chapter 9, and compare the T_{60}s we get with those obtained with the Sabine formula in Table 9-4. First we set up a six frequency tabulation, Table A, and observe the following steps:

Step 1-Enter the total sabins for each frequency from Table 9-4.

Step 2-Calculate the average absorption coefficient (a_{ave}) for each frequency by dividing the total sabins of absorption at each frequency by the total surface area of the room, 1620 sq ft.

Step 3 - We now enter the bottom of the graph of Figure A with the a_{ave} from Step 2, go up to the solid line curve and read the corresponding value of A from the left hand scale. Enter this value in Step 3 and repeat for the other frequencies.

Step 4-Now multiply the total surface area (S_T), 1620 sq ft in this case, by the value of A from Step 3. This gives a corrected value of total sabins.

Step 5 - Figure the T_{60} for each frequency in the usual way by dividing (0.05) (V) or 200 by the corrected total sabins from Step 4.

How different are these newly calculated Eyring T_{60}s from tthe Sabine T_{60}s calculated in Table 9-5? In Step 6 the T_{60} values from Table 9-4 have been recorded for ready comparison. We see that the new Erying values are 9 to 15 percent less than the Sabine values for this particular room. **The deader the room, the greater will be the difference.** We can see this from the graph of Figure A. The broken line represents a one-to-one relationship between a_{ave} and A. For low values of a_{ave} (very live rooms), the value of A is very close to a_{ave} and the T_{60}s calculated from the two formulas will be very close together. For high values of a_{ave} (very dead rooms) the value of A departs greatly from a_{ave} and the use of the Eyring formula is strongly urged.

For the usual listening rooms, radio, and recording studios have a reasonably high absorption, a value of

$$\frac{(0.027) \ (Volume)}{S_T}$$

can be subtracted from the Sabine formula to approximate what would be obtained with the Eyring formula. For this room this means subtracting 0.066 from the Sabine values of T_{60}. In Step 7 this subtraction has been carried out and the resulting values of T_{60} are very close to the Erying values in Step 5. This is fortuitous, but this lump correction works reasonably well for the narrow range of a_{ave} values encountered in conventional studios.

212

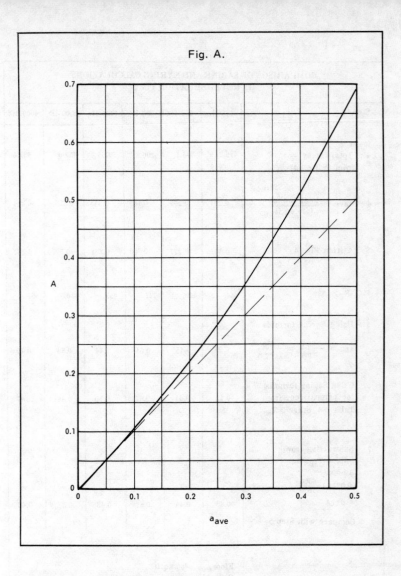

Fig. A.

COMPARISON OF SABINE AND EYRING CALCULATIONS
OF REVERBERATION TIME

Step		125 Hz	250 Hz	500 Hz	1000 Hz	2000 Hz	4000 Hz
1	Total sabins from Table 9-4	476.2	393.1	296.6	397.5	452.0	427.0
2	$a_{ave} = \dfrac{\text{total sabins}}{1620 \text{ sq ft}}$	0.294	0.242	0.183	0.245	0.279	0.264
3	A(from Fig. A)	0.348	0.277	0.202	0.281	0.327	0.307
4	$(S_T)(A)$	564.	448.	327.	455.	530.	497.
5	Using Eyring formula $T_{60} = \dfrac{0.05 \, V}{S_T A} = \dfrac{200}{S_T A}$	0.35	0.45	0.61	0.44	0.38	0.40
6	Using Sabine formula for comparison, from Table 9-4, Stage 3 T_{60}	0.42	0.51	0.67	0.50	0.44	0.47
7	Subtracting from Sabine T_{60}s an amount: $(0.027)\,(\dfrac{4000}{1620}$ $= 0.066$ Compare with Step 5	0.35	0.44	0.61	0.43	0.37	0.40

```
Floor........ 400 sq ft
Ceiling...... 400
N. Wall..... 160
S. Wall......160
E. Wall..... 250
W. Wall.....250

S_T = Total surface area.....1620 sq ft
```

REFERENCES

1. Olson, Harry F., "Psychology of Sound Reproduction," **Audio**, Vol. 56 No. 6, June 1972, pp 20-22, 24, 73.
2. Sprinkle, Melvin C., "Acoustics For Audio Men," **dB Magazine**
Part 1: Vol. 5 No. 3 March 1971, pp 21-23
Part 2: Vol. 5 No. 5 May 1971, pp 26-29
Part 3: Vol. 5 No. 6 June 1971, pp 21-25
3. Anderson, Roger and Robert Schulein, "A Distant Micing Technique," **dB Magazine**, Vol.5 No. 4, April 1971, pp 29-31
4. Stark, Craig, "The Sense of Hearing," **Stereo Review**, September 1969, pp 66, 71-74.
5. Stevens, S.S., Fred Warshofsky, and Editors of LIFE, "Sound and Hearing," **LIFE Science Library**, Time, Inc., N.Y., 1965.
6. Bergeyk, William A., John R. Pierce, and Edward E. David, Jr., "Waves and the Ear," Anchor Books, Doubleday (1960). This paperback is an authoritative, reasonably simple exposition of the functioning of the human ear.
7. Jacobs, Rudolph A., Jr., "The Loudness Control," **Electronics World**, December 1963, pp 34, 35, 82, 84. Elementary review of the subject which also contains a description of a simple circuit for loudness compensation.
8. "A Generator of Random Electrical Noise," General Radio Company, Engineering Department **Reprint No. E-110**. West Concord, Mass. 01781. In addition to describing the GR random noise generator, this reprint contains a discussion of the characteristics of random noise, the wide scope of its application, and a good bibliography on the subject. Figs. 3-6, 3-7, and 3-8 of this book are adapted from this paper.
9. Edwards, Lon, "White Noise—Its Nature, Generation and Applications," **Electronics World**, November 1962, pp 40-42.
10. For example, that manufactured by EMT (Electromesstechnik), West Germany.
11. Brüel, Per V., "Sound Insulation and Room Acoustics," Chapman and Hall, Ltd., London (1951). Gives the European approach to many practical studio problems. Some mathematical knowledge required for full understanding of many sections.
12. Architect, Aarno Ruusuvuori; Acoustics Consultant, Mauri Parjo.
13. Bolt, R.H., H. Feshback, and A.M. Clogston, "Perturbation of Sound Waves in Irregular Rooms," **Jour. Acoust. Soc. Am.**, Vol. 14, No. 56 (1942).
14. Knudsen, Vern O., "Resonances in Small Rooms," **Jour. Acoust. Soc. Am.**, Vol. 4, July 1932, pp 20-37.
15. Knudsen, V.O. and C.M. Harris, "Acoustical Designing in Architecture," Wiley, New York (1950). Chapter 8 has a nonmathematical treatment of room resonances and the book is recommended for those interested in all phases of room acoustics.

16. Brociner, Victor, "Loudspeakers—Can We Measure What We Hear?" **Electronics World**, March 1969, pp 25-29, 74, 75. A review of the problem of making measurements on loudspeakers which correlate with ear evaluations. A reverberant room technique is described which gives reasonable correlation with the way a loudspeaker sounds in the home.

17. Mankovsky, V.S., "Acoustics of Studios and Auditoria," Focal Press, Ltd., London (1971). The translation from the Russian has been edited by Dr. Christofer Gilford, an accomplished acoustician. Written as a textbook for university students, and presupposing mathematical proficiency, it fills in many holes left by other books on the subject.

18a. Jackson, G.M. and H.G. Leventhall, "The Acoustics of Domestic Rooms," **Applied Acoustics** (5) (1972).

18b. Gilford, C.L.S., "The Acoustic Design of Talks Studios and Listening Rooms," **Proc. Inst. of Elect. Engrs.**, Vol. 106 Part B No. 27, May 1959, pp 245-256. A helpful paper on the treatment of colorations of small rooms. The relative importance of axial, tangential, and oblique modes is discussed as well as conditions for their audibility. Practical considerations of room design dominate this paper.

19. See Reference 15.

20. Bolt, R.H.,"Note on Normal Frequency Statistics for Rectangular Rooms," **Jour. Acoust. Soc. Am.** Vol. 19 No. 1, July 1946, p 130.

21. Rettinger, Michael, "Acoustics—Room Design and Noise Control," Chemical Publishing Company, Inc., New York (1968). A useful reference book. Summary of room proportions is given in pp 241-244.

22. Sepmeyer, L.W., "Computed Frequency and Angular Distribution of the Normal Modes of Vibration in Rectangular Rooms," **Jour. Acoust. Soc. Am.** Vol. 37 No. 3, March 1965, pp 413-423.

23. Adapted from Reference 13.

24. Volkman, L.E., "Polycylindrical Diffusers in Room Acoustic Design," **Jour. Acoust. Soc. Am.** Vol. 13, January 1942, pp 234-243.

25. Boner, C.P., "Performance of Broadcast Studios Designed With Convex Surfaces of Plywood," **Jour. Acoust. Soc. Am.**, Vol. 13, January 1942, pp 244-247.

26. Everest, F. Alton, "The Acoustic Treatment of Three Small Studios," **Jour. Audio Engineering Society**, Vol. 16 No. 3, July 1968, pp 307-313.

27. Brown, Sandy, "Acoustic Design of Broadcasting Studios," **Jour. Sound and Vibration** (1964) I (3) pp 239-257. Paper describes BBC practice of that era.

28, 29. "A Guide To Selecting Concrete Masonry Walls For Noise Reduction," National Concrete Masonry Association (1970). Fig. 7-4 data on 4" concrete block wall from Solite Corporation, test KAL 359-1-66. Data on 4" concrete block wall plastered both sides also from Solite, test KAL 359-7-66. Tests were made by Kodaras Acoustical Laboratories. Fig. 7-5 data on plain 8" concrete block wall from Solite Corporation test KAL 359-3-66 made by Kodaras Acoustical Laboratories. The data on 8" concrete block wall plastered both sides from LECA, Norway, test on Wall C made by Acoustical Laboratory in Copenhagen, Denmark.

30. "Solutions To Noise Control Problems," Owens-Corning Fiberglas Corporation **Publication No. 1-BL-4589**. Aimed primarily at apartments, motels, and hotels, the information contained on isolating properties of various wall and floor structures is equally applicable to listening rooms and studios. Fig. 7-6 data on standard stud wall, Fig. 7-7 data on

staggered stud wall and Fig. 7-8 data on double stud wall all taken from this helpful manual.

31. "Performance Data—Architectural Acoustical Materials," Acoustical and Insulating Materials Association **Bulletin No. XXXI**, 1971-72. This compilation of absorption coefficients and other data on the products of ten American acoustical manufacturers is of great value in listening room and studio design. It is available at US $1.00 per copy through the Association at 205 West Touhy Avenue, Park Ridge, Illinois 60068.

32. Evans, E.J., nad E.N. Bazley, "Sound Absorbing Materials," National Physical Laboratories (1960). Crown copyright reserved. Available from Her Majesty's Stationery Office, London, or from Sales Section, British Information Services, 845 Third Avenue, New York, N.Y. 10022 for US $1.00. This publication is worth ordering and its value is not damaged by age.

33. Burd, A.N., C.L.S. Gilford, and N.F. Spring, "Data For The Acoustic Design of Studios,.. BBC Engineering Division **Monograph Number 64**, November 1966. A useful compilation available from BBC Publications, 35 Marylebone High Street, London W. 1 for 5s post free.

34. Reference 17, page 374.

35. Batchelder, J.H., W.S. Thayer, and T.J. Schultz, "Sound Absorption of Draperies," **Jour. Acoust. Soc. Am.**, Vol. 42 No. 3, September 1967, pp. 573-575.

36. Sabine, Paul E. and L.G. Ramer, "Absorption-Frequency Characteristics of Plywood Panels," **Jour. Acoust. Soc. Am.**, Vol. 20 No. 3, May 1948, pp. 267-270.

37. Brillouin, Jacques, "Sound Absorption By Structures With Perforated Panels," **Sound and Vibration** Vol. 2 No. 7, July 1968, pp. 6-22. This is a translation by T.J. Schultz of a classical paper in French. Although highly mathematical, this paper is application oriented and probably the most comprehensive treatment of the subject.

38. Rettinger, Michael, "Low-Frequency Sound Absorbers," **dB Magazine**, Vol. 4 No. 4, April 1970, pp. 44-46.

39. "Sound Absorption of Fiberglass PF Insulation," Fiberglass Standards **Supplement No. AC6.A2** published by Owens-Corning Fiberglass Corporation, April 1954.

40. Seikman, William, "Outdoor Acoustical Treatment: Grass and Trees," **Jour. Acoust. Soc. Am.**, Vol. 46 No. 4 (Part 1), October 1969.

41. Nixon, G.M., "Recording Studio 3A," **Broadcast News**, September 1947, pp 33-35.

42. Designed by Mr. Ian Campbell, Lecturer in Architecture, University of Hong Kong, with the collaboration of Mr. Gordon Bell, then Director of Far East Broadcasting Co., Inc., Hong Kong.

43. Jordan, Vilhelm L., "The Application of Helmholtz Resonators to Sound-Absorbing Structures," **Jour. Acoust. Soc. Am.**, Vol. 19 No. 6, November 1947, pp 972-981.

44. Snow, William B., "Recent Applications of Acoustical Engineering Principles In Studio and Review Rooms," **Jour. Soc. Mot. Pic. and Television Engrs.**, Vol. 70, January 1961, pp 33-38.

45. Davis, Don, "Calibrated Monitoring Systems," a reprint from **dB Magazine** available from Altec-Lansing, 1515 S. Manchester Avenue, Anaheim, California 92803.

46. Davis, Don and Don Palmquist, "Equalizing The Sound System To Match The Room," **Electronics World**, January 1970. Reprint available from Altec-Lansing (See Ref. 45).

47. Davis, Don, "Sound Systems Equalization," **Progressive Architecture**, September 1969. Reprint available from Altec-Lansing (See Ref. 45).

48. Davis, Don, "Facts And Fallacies on Detailed Sound System Equalization," **Audio Magazine**. Reprint available from Altec-Lansing (See Ref. 45).

49. Allison, Roy F., "The Loudspeaker / Living Room System," **Audio**, Vol. 55 No. 11, November 1971, pp. 18, 20, 22. A study of actual living room conditions by the Vice-President of Acoustic Research. He doubts the justification of narrow band living room equalization if the speaker systems are all good.

50. Thanks to Dr. Irwin A. Moon, Manager, Moody Institute of Science for supplying the graphic level recorder tapes taken in their studio C upon which this example is based.

51. Vlahos, Petro, "An Acoustic Response Standard," **Jour. Soc. Mot. Pic. and Television Engrs.**, Vol. 78 No. 12, December 1969, pp. 1043-1044.

52. Ljungberg, Lennart, "Standardized Sound Reproduction in Cinemas and Control Rooms," **Jour. Soc. Mot. Pic. and Television Engrs.**, Vol. 78 No. 12, December 1969, pp. 1046-1053.

53. Rasmussen, Erik, "A Report on Listening Characteristics in 25 Danish Cinemas," **Jour. Soc. Mot. Pic. and Television Engrs.**, Vol. 78 No. 12, December 1969, pp. 1054-1057.

54. Buckle, C.C. and A.W. Lumkin, "The Evaluation and Standardization of the Loudspeaker-Acoustics Link in Motion Picture Theaters," **Jour. Soc. Mot. Pic. and Television Engrs.**, Vol. 78 No. 12, December 1969, pp. 1058-1063.

55. If at all possible, professional help should be engaged for solution of your acoustical problems. Reputable acoustical consultants in your area may be located through acoustical materials suppliers, university physics departments or through members of the Acoustical Society of America. Inquires to the society may be addressed to American Institute of Physics, 355 East 45th Street, New York, N.Y. 10017.

56. Watters, B.G., "The Sound of a Bursting Red Balloon," **Sound**, Vol. 2 No. 2, March-April 1963, pp. 8-14.

57. Dobkin, Robert C., "Logarithmic Converters," **IEEE Spectrum**, Vol. 6 No. 11, November 1969, pp. 69-72.

58. Delpeck, Jean F., "Logarithmic Amplifier Has 66-dB Range," **Electronics**, October 17, 1966, pg. 89.

59. Spencer-Kennedy Laboratories Model 507 Decay Rate Meter. Address: 1320 Soldiers Field Rd., Boston, Mass. 02135.

60. Model RT60 Instant Reverberation Time Readout Device, Communications Company, 3490 Noell St., San Diego, California 92110.

61. Rudmose, Wayne, "Primer on Methods and Scales of Noise Measurements," **dB Magazine**: Part 1—Vol. 4 No. 1 January 1970, pp. 22-24; Part 2—Vol. 4 No. 2 February 1970, pp. 24-25; Part 3—Vol. 4 No. 3 March 1970, pp. 27-29; Part 4—Vol. 4 No. 4 April 1970, pp. 54-55.

62. Katz, Bernard, "A Primer on Sound Level Meters," **Audio**, Part 1 Vol. 53 No. 7, July 1969, pp. 22-24; Part 2 Vol. 53 No. 8, August 1969, pp. 42-44.

63. "Monosyllabic Word Intelligibility," American National Standard USAS S3.2-1960. Available from ANSI, 1430 Broadway, New York, N.Y. 10018.

MISCELLANEOUS REFERENCES

Misbett, Alec, "The Technique of the Sound Studio," Hastings House, New York, 10 East 40th Street, N.Y. 10016, 2nd edition revised and enlarged (1970), 599 pp., $13.50. This edition is a practical handbook of studio techniques rather than a technical work. It should be valuable to anyone working in and around studios and recording equipment.

Rettinger, Michael, "Sound Insulation For Rock Music Studios," dB Magazine, Vol. 5 No. 5, May 1971, pp. 30-32.

Spring, N.F. and K.E. Randall, "Permissible Bass Rise In Talks Studios," BBC Engineering, No. 83, July 1970, pp. 29-34. By "bass rise" is meant rise in bass reverberation time.

Volkmann, John E., "Acoustic Requirements of Stereo Recording Studios," Jour. Audio Engr. Soc., Vol. 14 No. 4, October 1966, pp. 324-327.

Graham, William R. "Studio Construction Techniques," dB Magazine, Vol. 4 No. 4, April 1970, pp. 34-37.

Hansen, Robert, "Studio Acoustics," dB Magazine, Vol. 5 No. 5, May 1971, pp. 16, 21-24.

Everst, F. Alton, "The Strange Antics of a Studiofull of Air," International Christian Broadcasters Bulletin, September 1971, pp. 7-10.

Schroeder, M.R., "New Method of Measuring Reverberation Time," Jour. Acoust. Soc. Am., Vol. 37 (1965) pp. 409-412. See also U.S. Patent No. 3,343,627.

Broch, J. T. and V. N. Jensen, "On The Measurement of Reverberation," Brüel & Kjaer Technical Review, No. 4 (1966).

Davis, Don, "Acoustical Tests and Measurements," Howard W. Sams & Co., Inc.

Rettinger, Michael, "Acoustic Design And Noise Control," Chemical Publishing Co., Inc., New York (1973). A new edition of Reference 21.

Gilford, Christofer, "Acoustics For Radio and Television Studios," Peter Peregrinus, Ltd., 2 Savoy Hill, London WC2R 0BP (1972). A summary of advances in studio acoustics since World War II.

INDEX

A

Absorber, low-peak	139
Absorber, resonant	102
Absorbers, acoustic	22
Absorbers, midrange	103
Absorbers, panel	94
Absorbers, perforated panel	99
Absorbers, porous	91
Absorbers, slat	102
Absorbers, sound	91
Absorption	72
Absorption, bass	97
Absorption coefficients	92,95,114
Absorption coefficient, sound	22
Absorption of sound	21
Absorption, peak	96
Absorption units (sabins)	114
Acoustic absorbers	22
Acoustic cortex (brain)	27
Acoustic design, studio	135
Acoustic flats	156
Acoustic response	165
Acoustics, bathroom	53
Acoustics, evaluating studio	173
Adjustable acoustics	154
Adjustable element, Snow	161
Air molecules	11
Airborne noise	79
Amplifier, log	179
Amplifier-loudspeaker	176
Amplitudes	8
Analyzer, ear as an	36
Analyzer, frequency	47
Anechoic chambers	163
Anvil	26
Articulation, percentage	187
Aural mechanism	25
Axial mode	59,64,119,135,144
Axial mode colorations	144,187

B

Band-rejection filters	169
Bandwidth limitation	50

Bass absorption	97
Bathroom acoustics	53
Binaural hearing	36
Binaural localization	36
Board, gypsum	83
Buzz	183

C

Calculation of T_{60}— the Sabine equation	113
Calculations, reverberation	57
Chamber, reverberation	21
Chambers, anechoic	163
Cochlea	26
Coefficients, absorption	92,95,114
Coefficient, sound absorption	22
Coherence	147
Coloration	58,60,66,113,167
Colorations, axial mode	144,187
Colorations, definition	60
Colorations, mode	142
Colorations, pinpointing room	186
Comparison of wall structure	82
Compression	11
Computing reverberation	108
Concave surfaces	71
Concrete block	82
Conducting medium	9
Continuous spectra	44
Control, 5-band	168
Control of interfering noise	77
Control room design	150
Control room treatment	150
Control, 10-band	168
Control, 24-band	164
Controlled reverberation	109
Controls, tone	167
Convex surfaces—the poly	71
Cortex (brain), acoustic	27
Cross-modulation test	51

D

Decay traces 177
Decibel 16,28,33
Design factors, general 152
Diaphragm 91
Diaphragm action, noise
transmitted by 79
Diffraction 20,36
Diffraction grating 72
Diffuser, polycylindrical 71
Diffusing elements 71
Diffusing surfaces 162
Diffusion 167,178
Diffusion of sound 66
Diffusion problems 67
Diffusion, sound 175
Discrimination ability of
the ear 35
Distortion 38,50
Distortion, dynamic 50
Distortion in time 50
Distortion limit, acceptable 43
Distortion, nonlinear 51
Distortion, transient 51
Distribution of sound
energy 41
Divergence, geometrical 15
Doors, sound-insulating 88
Double windows 86
Draperies 155
Dynamic distortion 50
Dynamic range 41

E

Ear, discrimination
ability 35
Ear's frequency response 30
Ears—golden and tin 24
Earthquake 52
Echo, flutter 175,186
Elastic media 9
Elasticity 9
Electroacoustic coupling,
two rooms 134
Elements, diffusing 71
Elements, rotating 156
Equal-loudness contours,
Fletcher-Munson 28
Equalization limitations 167
Equalizing procedure for
home hi-fi 171
Evaluating speech
intelligibility 187
Evaluating studio acoustics 173

Evaluation of background
noise 183
Eyring reverberation
formula 133

F

Fader 35
Fan noise 184
Feedback howling 169
Fiber, mineral 81
Filter 176,185
Filter, scratch 50
Filters, band-rejection 169
Filters, passive 167
Five-band control 168
Flat response 163
Flats, acoustic 156
Fletcher-Munson contours 28
Flexure 91
Flutter 143,147,167
Flutter decay 186
Flutter echo 175,186
Flutter mode 186
Frequency 8,14
Frequency analyzer 47
Frequency and wavelength 14
Frequency, natural 55
Frequency response of
ear 30
Fundamental 8

G

General design factors 152
Geometrical divergence 15
Glass wool 99
Golden ears—and tin 24
Graphic level recorder 179
Grating, diffraction 72
Gypsum board 83

H

Hammer 26
Haney 182
Harmonics 8,44,60
Hear, how we 25
Hearing, binaural 36
Hearing, human 24
Hearing, limits of 29
Hearing threshold 29
Hertz 14
Hinged panels 158
Home hi-fi equalizing
procedure 170

House curve 165
How sound acts 7
Howling, feedback 169
Hum 183
Human hearing 24

I

Impulsive noise 176
In phase 19
Inertia 9
Intelligibility 111
Intelligibility, evaluating
speech 187
Intensity 8
Interference 19
Interfering noise,
control of 77
Inverse square law 15

L

Limits of hearing 29
Line spectra 43
Line spectrum 44
Listening chain 163
Listening problem 163
Liveness 97
Living room T_{60} 112
Localization 36
Localization, binaural 36
Log amplifier 179
Logarithmic relationship 27
Longitudinal waves 11
Loss, transmission 78
Loudness 8
Loudness level 31
Louvered panels 160
Low-peak absorber 139

M

Main studio design 135
Mass law graphs 81
Materials, placement of 106
Materials, porous 81
Microbar 33
Microphone 176
Mid-peak absorbers 140
Midrange absorbers 103
Mineral fiber 81
Modal frequency 58
Mode 58
Mode colorations 142
Mode colorations, axial 144
Mode spacing 60

Modules, 104,140,146
Molecules, air 11
Monaural 37
Music 40
Music, speech, noise 38

N

Natural frequency 55
Networks, weighting 185
Noise, airborne 79
Noise and room resonances 90
Noise, background,
evaluation 183
Noise carried by structure 77
Noise, fan 184
Noise, impulsive 176
Noise, interfering, control
of 77
Noise level, thermal 43
Noise, pink 49
Noise, random 47
Noise sources 77
Noise, speech, music 38
Noise—the bad kind 47
Noise—the good kind 47
Noise transmitted by
diaphragm action 79
Noise, white 49
Nonparallel walls 70
Nonlinear distortion 51
Normal modes and
reverberation 109

O

Oblique waves 59
Optimum reverberation
time 111
Organ of Corti 26
Oscillation 9
Oscilloscope 179
Ossicles 26
Oval window 26
Overhead feature 143
Overtones 44

P

Pain threshold 29
Panel absorbers 94
Panel absorbers,
perforated 99
Panels, hinged 158
Panels, louvered 160
Panels, portable 156

Panels, wraparound: polys 96
Particles 10
Passive filters 167
Patterns, standing wave 58
Peak absorption 96
Perforated panel absorbers 99
Periodic waves 44
Phase 8,36
Phase opposition 19
Phase relationships 45
Phase shifts 46
Phon 28
Phonemes 36
Pinpointing room
    colorations 186
Pipe, resonance in a 54
Pitch 8
Placement of materials 106
Plane surfaces 75
Poly—convex surfaces 71
Polycylindrical diffuser 71
Polys 127
Polys: wraparound panels 96
Porosity 91
Porous absorbers 91
Porous materials 81
Portable panels 156
Proportions, room 66
PVC 89

R

Random noise 47
Range, dynamic 41
Rarefaction 11
Ratio, S∕N 176
Ratio, signal-to-noise 39
Reflection of sound 17
Reradiation 72
Resonance 82
Resonance in a pipe 54
Resonances 52
Resonances in listening
    rooms and small studios 56
Resonances, room, noise and 90
Resonant absorber 102
Resonant devices, variable 159
Resonators 52
Response, acoustic 165
Response, flat 163
Response, nonuniform 50
Reverb time: stopwatch
    method 180
Reverberation 66,108,167
Reverberation and normal
    modes 109
Reverberation calculations 57
Reverberation chamber 21

Reverberation characteristic 154
Reverberation computation 108
Reverberation, controlled 109
Reverberation
    time 78,141,146,151,173
Reverberation, time
    optimum 111
Reverberation time—T_{60} 110
Rockwool 81
Room furnishings 120
Room proportions 66,119
Room resonances and
    noise 90
Rotating elements 156

S

Sabine 180
Sabine equation—calculation
    of T_{60} 113
Sabins (absorption units) 114
Scratch filter 50
Sensation 7
Signal sources 175
Signal-to-noise ratio 39,176
Sine wave 13
Slat absorbers 102
SLM 184
Snow adjustable element 161
"Soft" studio 133
Solutions to noise 77
Sone 28
Sound absorbers 91
Sound absorption 21
Sound absorption
    coefficient 22
Sound diffusion 175
Sound energy distribution 41
Sound-insulating doors 88
Sound-insulating walls 80
Sound level meter 184
Sound locks 88
Sound pressure level 34,182
Sound quality 8
Sound reflection 17
Sound spectrograph 40
Sound superposition 18
Sound transmission
    classification (STC) 81
Sound wave, how it moves 11
Spectograph, sound 40
Spectra, continuous 44
Spectra, line 43
Spectrum, line 44
Speech 38
Speech, music, and noise 38

Speech studio 146
Speech studio design 143
SPL 34
Standing wave 56,58,60,63
Standing wave patterns 58
STC 81
STC contour 82
Stimulus 7
Stirrup 26
Studio acoustics,
  evaluating 173
Studio, "soft" 133
Superposition of sound 18
Surfaces, concave 71
Surfaces, convex—the poly 71
Surfaces, diffusing 162
Surfaces, plane 75

T

Tangential waves 59
Ten-band control 168
Thermal noise level 43
Threshold of feeling 59
Threshold of hearing 29
Threshold of pain 29
Timbre 41
Time distortion 50
Tone controls 167
Transient distortion 51
Transmission loss 78
Treble T_{60} 113

T_{60} calculation, living
  room 121
Tuning the listening
  room 163
Twenty-four-band control 169
Two rooms coupled
  electroacoustically 134

V

Variable resonant devices 159
Velour 93
Vibration 9
Voiceprints 40
von Helmholtz 52

W

Wall structures,
  comparison of 82
Wall treatment 140
Walls, sound-insulating 80
Waveform 8
Wavelength 14
Wavelength and frequency 14
Wave, standing 56
Waves, longitudinal 11
Waves, periodic 44
Weighting networks 185
White and pink noise 49
Window, oval (in ear) 26
Windows, double 86
Wraparound panels: polys 96